geo
all about angles

Exploring Line and Angle

ALLAN TURTON
CALVIN IRONS

GEO All About Angles

contents

introduction

Geometry

Geometry is the exploration of space, size, and position. Our knowledge or intuitive understanding of geometry plays a major part in our everyday and professional lives. For example, a simple task of reading a map requires an understanding of direction and position. A more demanding task of repairing an engine draws on an intuition about shapes and the effect of moving them. The mechanic is required to visualize the shapes and how they fit together. Likewise scientists, engineers, and carpenters are all required to know and use certain geometric ideas that are specific to their profession. For this reason alone, it is easy to see why geometry has become an essential part of the school mathematics curriculum. However, there is an even more important reason for placing greater emphasis on the teaching of geometry. Geometry involves the manipulation of mental pictures, which is often called visual thinking. Problem solving in all strands of mathematics depends on forming mental pictures of the situation in which the problem is embedded and then "finding" a picture of the mathematical idea that matches. The ability to mentally form, rearrange, and match pictures is crucial to all aspects of mathematics, particularly problem solving.

The *GEO* Series

These *GEO* books are designed to help teachers make geometry an essential component of their mathematics curriculum. The eight books provide hundreds of exciting and innovative activities for teaching geometry to students aged 6 to 12 years and beyond. The topics include two-dimensional shapes; three-dimensional shapes; lines and angles; symmetry; tessellation; topology; folding two-dimensional shapes; and location, direction, and movement.

Each book contains easy-to-read instructions and fully reproducible blackline masters. Each activity has two types of icon: one indicates the appropriate age group for the activity and the other indicates whether the activity is designed as a whole class, small group, pairs, or individual activity. Definitions of terms are provided as friendly reminders and interesting facts appear in the margin. A glossary is also provided at the back of each book.

Those teachers who use all eight books will develop a greater understanding of geometry and new insights into its use. Selecting and using activities from the series will offer students a broad view of geometry and one that is inviting, engaging, and thought-provoking.

The term "geometry" simply means "earth measure".

▲ *These icons show the appropriate age group for the activity.*

▲ *These icons show whether the activity is designed for an individual, a pair, a small group or a whole class.*

About *GEO All About Angles*

This book contains a variety of exciting units which explore line and angle. It is divided into two sections to make it easier to choose activities that deal with only lines or only angles. However, to explore angles, lines need to be considered and to explore relationships between certain lines, angles need to be considered. So in some activities knowledge of both line and angle is required and may have to be introduced or revisited.

Activities in the first two units of this book investigate line. The following units contain activities that explore different types of angles and how to measure, compare, and calculate their sizes. Two of these units are sequenced to develop understanding of angle and angle measurement. The unit "Introducing Angles" is a starting point for activities with angles for students of all ages. It provides information on basic terminology and concepts for investigating angles. Similarly, the unit "Measuring Angles" provides a full sequence of lessons to develop ideas about how the size of angles can be described.

Assessment

Activities in the *GEO* series have been designed to help students demonstrate key outcomes and standards in geometry. An assessment chart has been designed to identify those activities that can help you assess your students' understanding and achievement. The chart is available from the following websites:

North America — www.origomath.com

Australasia — www.origo.com.au

Terminology

Emphasis is often given to remembering different terms for particular lines and angles. For example, a "line" technically extends infinitely into space, while a "line segment" is just a part of that line. A "ray" is similar to a line except that it has a starting point from which a line then extends infinitely into space. In reality however, these concepts can only be represented on paper by a line segment. Additionally, in everyday speech, all three concepts are generally called "lines". Because of this, and for ease of reading, this book uses "line" to describe lines, line segments, *and* rays. Teachers may like to model the correct terminology for their students (see Activity 12 on page 7), but it is not vital for students to remember it.

▲ *You may wish to discuss the mathematical concepts of lines, line segments, and rays with your students.*

Some types of angles have names that students find confusing and can lead to mistaken ideas about angles. For example, the term "right angle" suggests to some students that one of the two angle arms must point to the right and that, consequently, "left angles" may exist. The "Comparing Angles" unit provides opportunities for the students to discover where such terms originated and how they can be useful. To assist students in discussing and measuring angles of different sizes in an easier and more logical manner, reference is made to the fraction of a full turn that an angle represents. For example, a "quarter angle" is a quarter of a full turn, which means that a "half angle" is twice the size of a quarter angle, while an "eighth angle" is half the size.

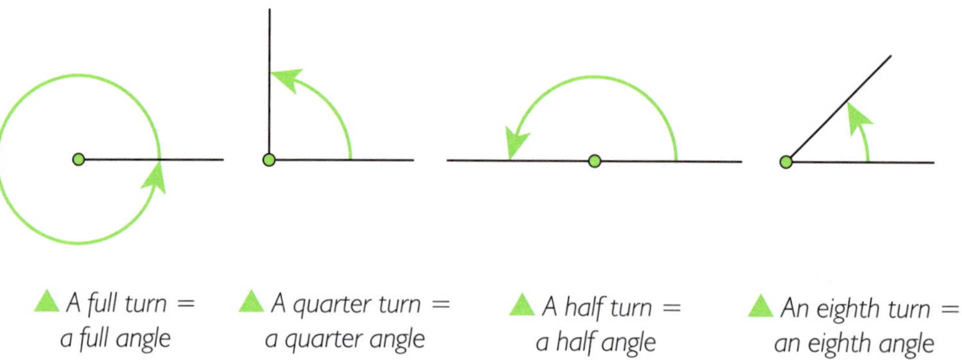

▲ A full turn =
a full angle

▲ A quarter turn =
a quarter angle

▲ A half turn =
a half angle

▲ An eighth turn =
an eighth angle

Most terms used to describe geometric shapes are the same in all English-speaking countries. However, some terms common in one place are not in another. One example involves naming types of rectangles. Some students find it confusing that squares are described as types of rectangles. The term "non-square rectangle" is then sometimes used to describe those rectangles that are not squares. A more concise way of describing such shapes is to call them "oblongs". Using this term, an analogy can be drawn between squares and oblongs being types of rectangles in the same way that a sedan and a limousine are both types of cars. For these reasons, the term "oblong" is used in this book.

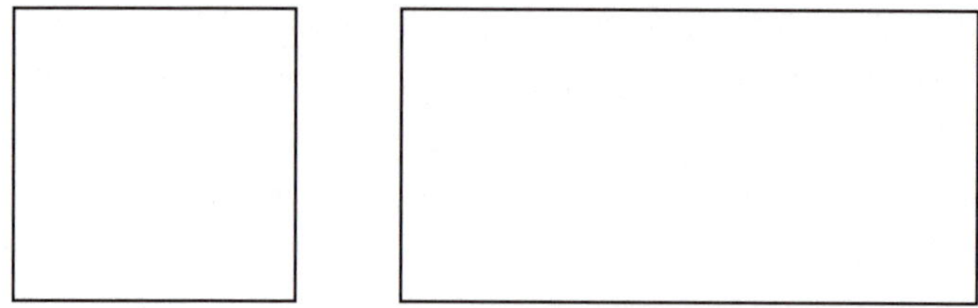

▲ Both of these shapes are rectangles, but in this book the shape
on the right is called an oblong.

Lines

As mentioned previously, lines are often separated into the related concepts of "lines", "line segments", and "rays" (among others). Additionally, lines are often described as having only one dimension, or "one-dimensional". There are several ways to define dimensions in order to discuss such a concept.

1. One use of "dimensions" is to describe a shape as having certain "properties". In this sense, a cube has three dimensions (length, width, and height), a square has two (length and width), and a line segment has only one (length). The difficulty with this definition is that one- and two-dimensional shapes can be described in words but not accurately drawn with pictures or made with materials. Using "dimensions" as a type of property to describe a line would mean that a "one-dimensional" line segment constructed from a piece of string itself would actually be "three-dimensional" because the string has length, width, and height. So using "dimensions" in this way can be confusing.

2. The word "dimensions" is also used it to describe the lineal measurements we often use to define a shape. In this example, "dimensions" are not properties of a shape, but measurements in a particular direction. Using this definition of "dimensions", a prism is measured using three dimensions (length, width, and height). A triangle, square, and even a circle are measured using two dimensions (length and width or similar), while a line segment is measured using only one dimension (length).

3. The most precise approach in using "dimensions" as measurement is to consider how many dimensions, or particular measurements, are needed to identify the position of a point on or inside any geometric figure. To locate a point anywhere along a line segment, all that is required is to know the distance from one end of the line segment to where the point is. If we think of a triangle lying on a co-ordinate grid, then two dimensions can be used to describe a point anywhere on or inside the triangle. Similarly, any point on or inside a cube can be found using three measurements from, for example, one of the corners on the base of the cube.

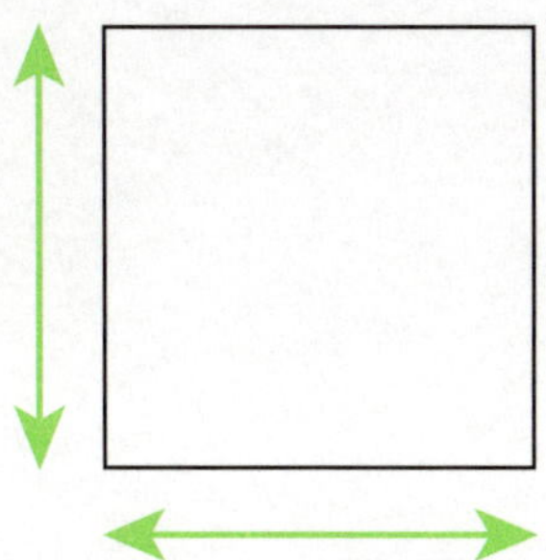

▲ A square is measured using two dimensions that happen to have the same value.

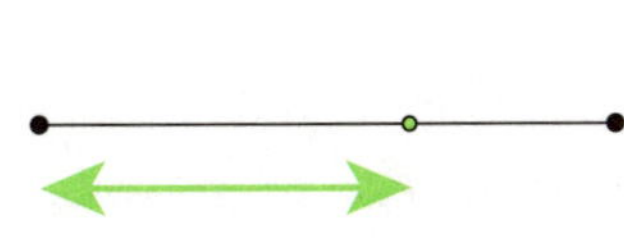

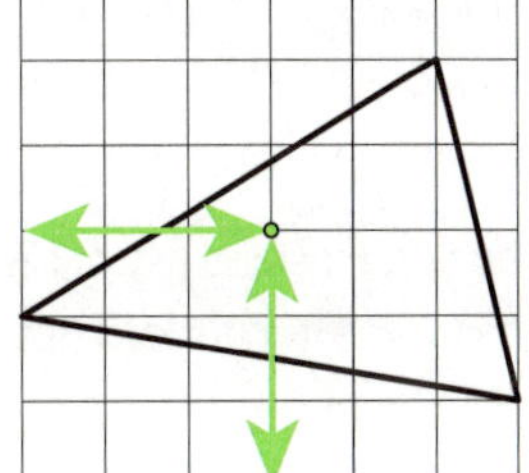

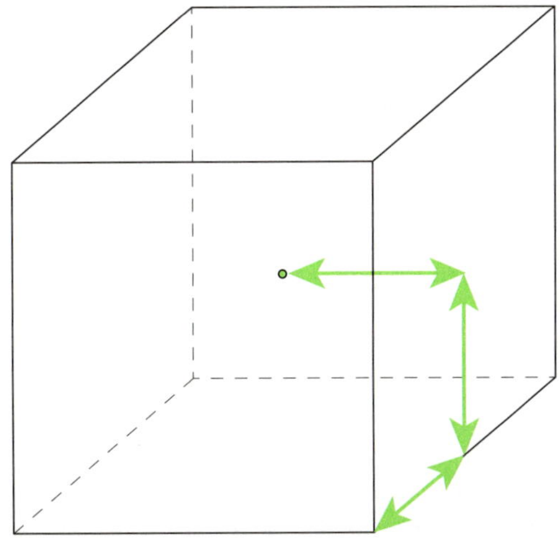

▲ Different numbers of dimensions are needed to describe the position of a point on these objects.

In summary, any concrete model of a one-dimensional shape will have length, width, and height. Students may better understand the concepts involved with this area of geometry if they consider dimensions as measurements and think of a one-dimensional shape as being a shape for which only one dimension is required to locate a point on the shape.

Angles

Some definitions of an angle suggest that an angle is a figure or object. For example, "an angle is the figure formed by two rays that share a common endpoint." Such definitions can lead some students to the incorrect conclusion that the longer the rays, the bigger the angle. More complicated issues arise when trying to use such a definition to describe a real-world situation like a turning door knob. The amount that a door knob turns can be measured in degrees but a "figure formed by two rays" cannot be seen at all.

A more useful approach to defining an angle is to describe it as an "opening" or "amount of turn" between two angle arms. Separating the ideas in such a definition produces these key features:

- there are two angle arms, whether real or imaginary
- there is a point where the two arms meet
- there is an amount of turn between the two arms

In some contexts the angle arms are fixed and in others they move. For example, the hour hand and the minute hand of a clock act as angle arms and show a variety of different angles. It is very easy to relate the concept of an angle being an amount of turn to situations such as this. With a bit more thought, it can also be seen that the front and back covers of a book are angle arms, with the spine acting as a stretched-out turning point that they are joined to.

When the angle arms are fixed it is initially harder to relate such situations to the "angle as amount of turn" concept. In a triangle the two angle arms may be two sides of the triangle and because the arms do not move, the amount of turn between them may be more easily described as an "amount of opening". Similarly, two adjacent faces of a cube act as fixed angle arms while the space between them has a certain amount of opening. However, care needs to be taken when discussing angles as "openings" that students do not come to the mistaken conclusion that the size of an angle is connected to the amount of area or volume between the angle arms.

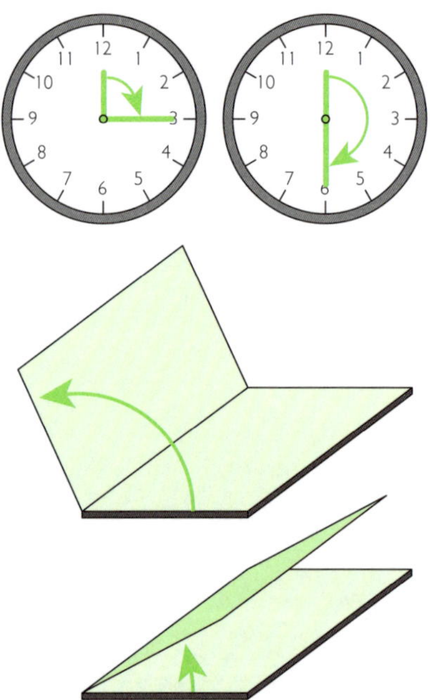

▲ When angle arms are moveable, it is easy to see how an angle can be described as an amount of turn.

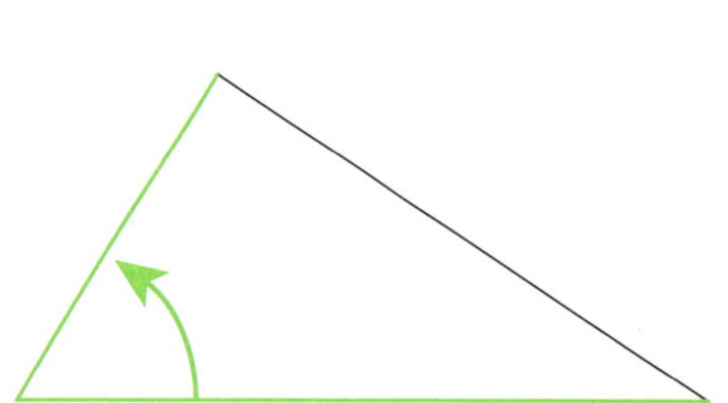

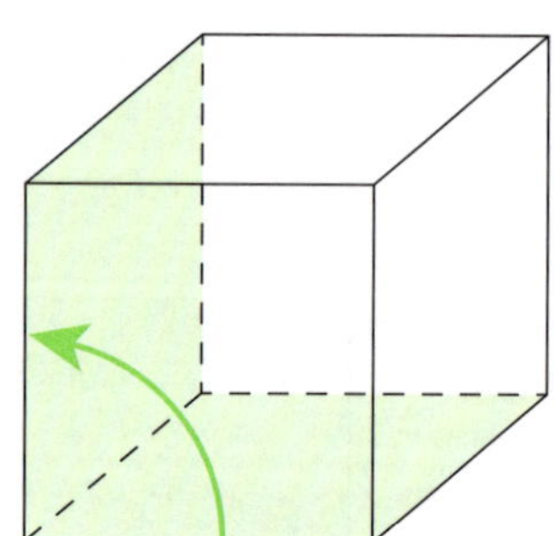

▲ When angle arms are fixed, it is easier to describe an angle as an amount of "opening" between the arms.

describing lines

In these activities, students explore the basic properties of different types of lines. As mentioned in the introduction to this book, the term "lines" is used in its everyday sense to describe the mathematical concepts of lines, line segments, and rays. If you wish to use the mathematically correct terminology with your students, Activity 12 provides an introduction to the concepts involved.

1. Word Wonders

Preparation

Make three copies of Blackline Master 1 onto tagboard or light card for each group of students and cut the letter cards to size.

■ Materials
- Blackline Master 1 (page 59)

Activity

The students play this game in groups of two to four. The game proceeds in rounds. In each round, each player has to make a word from any of the letters. The players gain points depending on the types of lines in each letter of their word. Every letter that has only straight line segments is worth one point, letters with only curved line segments are worth two points, and letters that have curved and straight line segments are worth three points. The example below shows the points earned for the word "bounce".

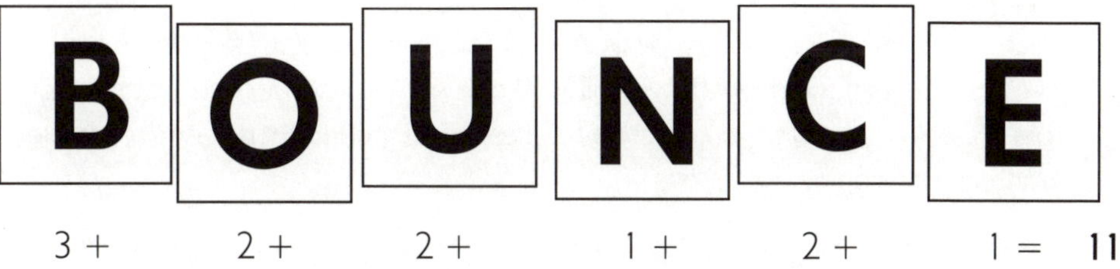

B	O	U	N	C	E
3 +	2 +	2 +	1 +	2 +	1 = 11

▲ *Playing this game is a fun way for students to identify straight and curved lines.*

2. Crossed and Separate

Preparation
No preparation is required.

■ Materials
- Drinking straws or skewers

Activity

1. Arrange the students in a circle and drop the straws or skewers into the middle of the circle. Ask the students to describe how the straws are positioned. Bring out the fact that some are lying on top of each other so that they are crossed over, while others have fallen separately.

2. Invite volunteers to identify pairs of straws that either cross over each other or are separate. Remove the straws once they have been identified. Repeat the process so that all students have a turn and until all of the straws are gone.

6-8

■ Materials

- Skipping rope — 1 for each pair of students

3. Making Waves

Preparation

No preparation is required.

Activity

Ask each pair of students to lay their rope flat on the ground and to arrange it to make straight or "wavy" lines. Discuss whether the curved ropes are "very wavy" or "almost straight". As a conclusion to the activity, invite the class to lay their ropes side by side to make "waves".

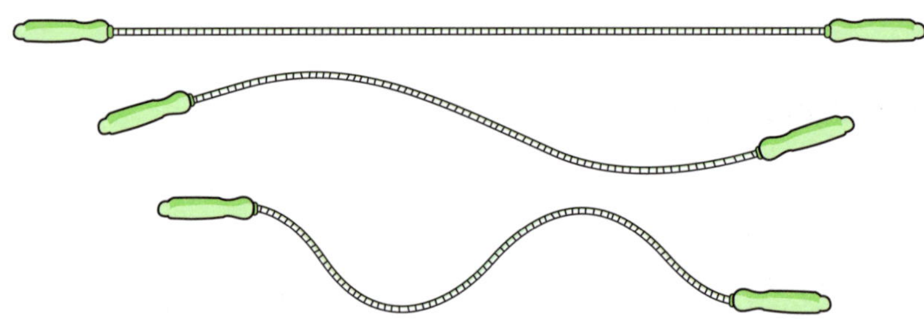

▲ *Using skipping ropes is a hands-on way for students to explore different lines.*

6-8

■ Materials

- Tennis ball
- Small marble — 1 for each pair or group of students
- Adhesive tape

Did You Know?

The word "horizontal" comes from "horizon", which can be defined as the line that is seen to separate the ocean from the sky when looking out to sea. This is reflected in the Greek origins of horizon, *horizein*, which means "to divide or separate". Any line that is parallel to the horizon is said to be "horizontal".

4. Horizontal Lines

Preparation

No preparation is required.

Activity

1. Place the tennis ball on a level desk. Ask, *What do you think will happen to the ball if I lift one edge of the desk?* (It will roll.) Lift one edge of the desk so that the ball rolls off. Ask, *How could you describe this desk?* The students may use words like "slanting" or "sloping".

2. Release the desk so that it is level again. Ask, *How could you describe the desk now?* The students may suggest words such as "level" or "flat". Invite the students to suggest words or expressions that use the concept of "level", for example "an even keel", "on the level", "a level tablespoon", or "sea level".

3. Show the students how to tape two pencils together to form a gutter that a marble can roll along. By placing the pencils and marble on a surface, it can be used as a tool to determine whether a surface is truly level or not.

▲ *Challenge the students to use this simple tool to find level surfaces or lines.*

4. Have the students work in groups of two or three to make level testers and to find examples of level (and non-level) surfaces in the classroom. Discuss the students' findings and introduce the word "horizontal" to describe something that is level.

5. Plumb Lines

Preparation
Attach a sinker to one end of the fishing line.

Activity

1. Holding the sinker in one hand, and the end of the fishing line in the other, raise both your hands so that they are level. Ask, *What do you think will happen if I let go of the weight at the end of this line?* Bring out the idea that not only will the weight swing down, then back and forth, but that, eventually, it will stop and the line will hang straight down.

2. Introduce the term "plumb line" and show how it can be held near straight edges to test whether they are "straight up and down". Have the students work in twos or threes using the fishing line and sinkers to make a plumb line. Direct them to find instances in the classroom where objects or edges are "straight up and down".

3. Afterward, invite volunteers to identify the examples they found and discuss the term "vertical". Introduce the term "oblique" to describe lines or surfaces that are neither horizontal nor vertical.

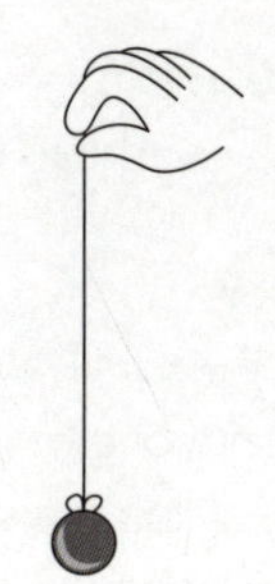

▲ *The students can use a plumb line made from fishing line and sinkers to find vertical lines or surfaces.*

■ Materials
- Fishing line or string — 1 length for each pair or group of students and 1 for demonstration
- Lead sinkers or similar (e.g. steel washers or nuts) — 1 set for each pair or group of students and 1 for demonstration

Did You Know?
The term "plumb line" comes from the Latin word for lead (the metal), *plumbum*. A plumb bob, the weight at the end of a plumb line, used to be made from lead. A related word is "plumber"; a person who traditionally worked with lead pipe.

---

6. Changing Horizons

Preparation
Draw a straight line that is parallel to the long sides of the paper.

Activity

1. Show the students the sheet of paper. Discuss using the "top" or "bottom" edges of sheets of paper as substitute horizons. The "sides" of the paper are used as reference points to determine whether a line segment drawn on the paper is vertical. Ask, *How would you describe the line that is drawn on this sheet of paper?* (Vertical.)

2. Rotate the paper slightly. Ask, *Is the line still vertical?* The answers to this question will vary. In an absolute sense, the line is not vertical, as can be seen by holding a plumb line against the paper (see the previous activity). In a relative sense, if the edges of the paper are references for what is vertical and horizontal, then the line is vertical.

3. Continue to rotate the paper. Explore with the students how far a piece of paper can be turned before a line segment appears closer to being horizontal than vertical. Conclude by saying that, in everyday use, the top edge of a piece of paper is used as a horizon and that the "top" is determined by how the words are positioned on the page.

■ Materials
- Large sheet of paper

Did You Know?
The word "perpendicular" originally described a plumb line, and the way it would hang downward to make a quarter angle with horizontal ground. "Perpendicular" came to refer to any two straight lines or flat surfaces that formed a quarter angle together. A word with similar roots is "pendant", an object that also hangs downward.

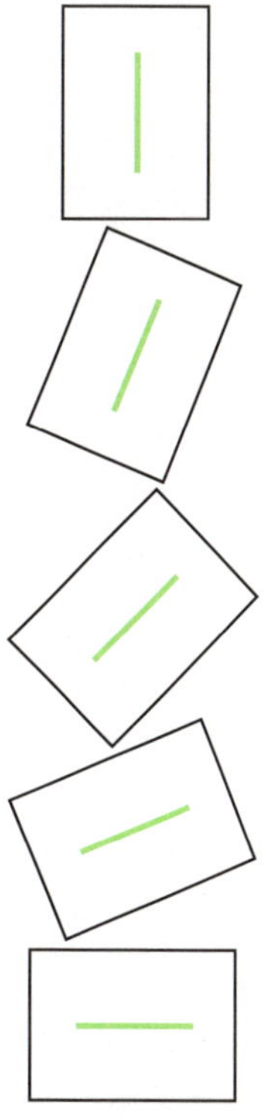

▲ *Discuss how to describe a line that has been turned from a vertical position.*

■ Materials

- Blackline Master 2 (page 60)
- Overhead projector and blank transparency sheet

Did You Know?

An oblique is a line or surface that is neither vertical nor horizontal.

A "vertex" is the point where two or more line segments join or intersect on a 2D shape, where three or more edges meet on a 3D shape, or the point where two or more angle arms intersect.

■ Materials

- Sheet of paper — 1 for each student and 1 for demonstration
- Hardcover book

Did You Know?

A quarter angle is one-quarter of a complete turn around a point. It is equal to 90°.

When two lines or surfaces intersect to form a right angle they are said to be "perpendicular" to each other.

7. Diagonal Lines

Preparation

Make an overhead transparency (OHT) of Blackline Master 2.

Activity

Display the OHT and direct each student to talk with a partner about the question at the bottom of the sheet. Invite students to share their ideas, and bring out the fact that a diagonal line is a straight line that joins opposite (or non-adjacent) vertices inside a shape. Invite the students to identify the diagonal line segments in figures such as those shown below. Students will see that some lines are oblique *and* diagonal.

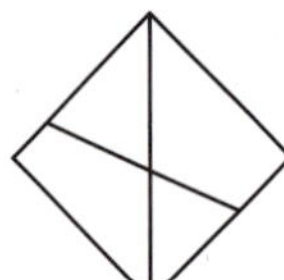 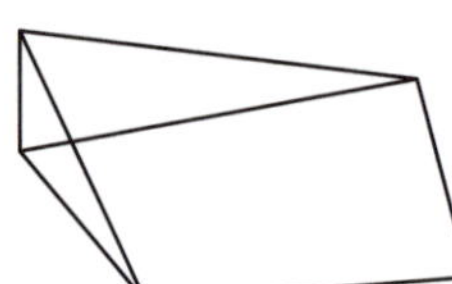 

▲ *Invite the students to identify oblique lines and diagonal lines in figures such as those shown above.*

8. Exploring Parallel Lines

Preparation

Make a quarter-angle tester by folding a sheet of paper in half, then in half again.

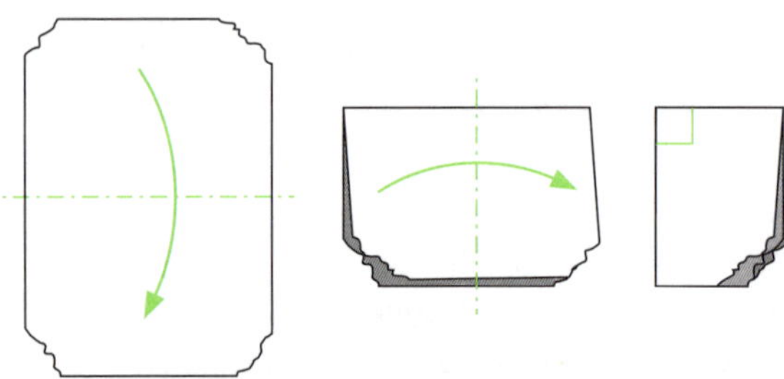

Activity

1. Show the students the cover of the book. Indicate the top edge and the bottom edge. Say, *I think these two edges are the same distance apart along their entire length. How can I prove it?* (Use a ruler to measure.)

2. Have a student hold the book so that the class can see the cover. Place a ruler on the book cover so that it is aligned with a side of the book, and measure the distance between the top and bottom edges. Read out the distance, then measure it by placing the ruler obliquely on the cover. Read out the measurement. It will be different from the first measurement. Ask, *Can these two edges really be the same distance apart along their entire lengths? What did I do wrong?*

3. Say, *When we measure the distance between two lines to see if they are parallel, we have to make sure the edge of the ruler is perpendicular (makes a quarter angle) with one of the lines. What could I use to help me make sure the edge of the ruler is perpendicular with the bottom edge of the book cover?* (A quarter-angle tester or protractor.)

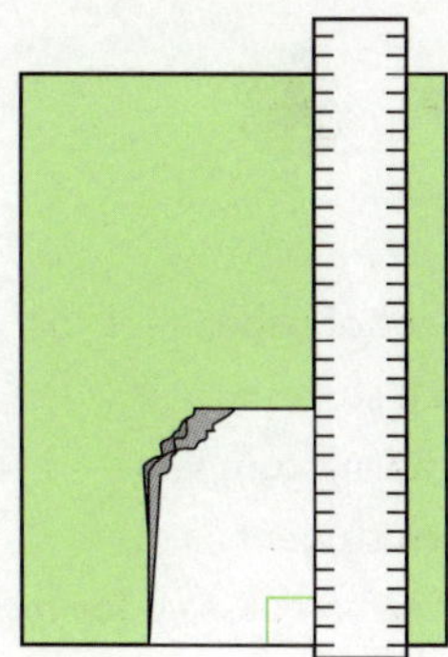

▲ *The pictures on the left and right show the correct way to measure the distance between parallel lines. Ask the students to explain why the middle method will not work.*

▲ *Students can use a quarter-angle tester to ensure they have their rulers positioned correctly to measure the distance between parallel lines.*

Demonstrate the proper technique as shown above, then invite the students to estimate which objects in the room have parallel lines, before testing with a ruler and quarter-angle tester.

9. Diagonals in Quadrilaterals

10–12

Preparation

Make a copy of Blackline Master 3 for each student.

Materials

- Blackline Master 3 (page 61)

Activity

1. Discuss the shapes shown on Blackline Master 3. Focus on the properties of the shapes.

2. Direct the students to draw two diagonal lines within each quadrilateral. Say, *The diagonal lines in each of these shapes have special properties. These special properties might be something to do with the length of the lines or the angles that they make with each other.* Encourage the students to work in pairs to examine the shapes. If any students need assistance to begin, ask them to note where the diagonal lines intersect each other in the parallelogram. Bring out the fact that each diagonal bisects the other.

3. As the students investigate the shapes, have them use the bottom of the grid paper to test whether the relationships they notice are true for different-sized examples of the quadrilaterals shown at the top of the paper.

4. Afterward, discuss the students' results. By comparing all of their drawings, the following features should be discovered:

   - *The diagonals of a parallelogram bisect each other.*
   - *The diagonals of a square bisect each other. They also form quarter angles with each other.*
   - *The diagonals of a rectangle bisect each other.*
   - *The diagonals of a rhombus bisect each other. They also form quarter angles with each other.*
   - *The diagonals of a kite bisect and form quarter angles with each other.*

Did You Know?

Sometimes diagonal lines are incorrectly described as being "equivalent to an oblique line". Diagonal lines may actually be vertical, horizontal, or any other orientation. If a diagonal line does happen to lie obliquely it is described as an "oblique diagonal line".

Did You Know?

To bisect a line segment or angle is to divide in it half.

geo For dozens of interesting activities on exploring the properties of quadrilaterals, see *Paper Polygons* and *Plane Puzzles*.

Materials

- Sheet of paper — 1 for each student
- Drawing compass — 1 for each student
- Large compass — for the board

10. Curved Circle Lines

Preparation

No preparation is required.

Activity

1. Draw a large circle on the board.

2. Instruct the students to mark a small point. Explain that the point will be the center of a circle. Direct the students to move the compass legs to 5 cm apart. Tell them to place the compass needle on the center point they marked and draw a circle.

3. Say, *It can be a little confusing to describe a circle. Some people think that a circle is the curved line that we just drew. Others think that it includes the area inside the line. One word to describe just the curved line is "circumference".*

4. Direct the students to mark any two points on the circumference of their circles. Say, *A section of a circumference is called an "arc". Whenever you have one arc on a circumference, there is always another one to complete the circle. Usually, there is a large arc and a small arc. Sometimes there are two equal arcs. How do you think that could happen?*

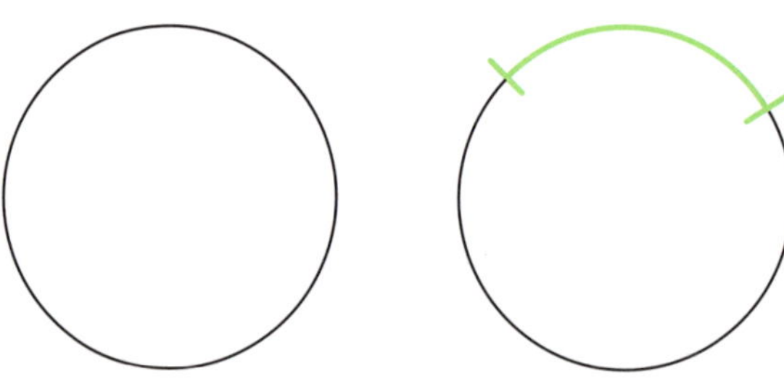

▲ *By marking two points on a circumference, the students will see that there are two arcs.*

Materials

- Sheet of paper — 1 for each student
- Drawing compass — 1 for each student
- Large compass — for the board

11. Straight Circle Lines

Preparation

No preparation is required.

Activity

1. Following on from the previous activity, ask the students to draw a second circle. Direct them to draw a straight line from the center of the circle to the circumference. Introduce the word "radius" to describe this line. Have the students draw another radius and introduce the plural of radius as "radii". Say, *Measure the lengths of the two radii. What do you notice?* (They are the same length.)

2. Instruct each student to mark any two points on the circumference of the circle and to draw a straight line to join them. Introduce the name for this line as a "chord". Have each student draw a second chord, measure both chords, and compare the results with a partner. The students will determine that two chords in a circle will only be the same length by coincidence.

3. Ask, *How can we make sure that two chords in a circle are exactly the same length?* Discuss the students' thoughts and bring out the idea of drawing a chord through the center of the circle. Have each student try this a few times to verify that these chords are exactly the same length. Say, *A chord that goes through the center of the circle is called a "diameter". What do you notice about the two arcs that are formed by the end points of a diameter?* (They are the same size.)

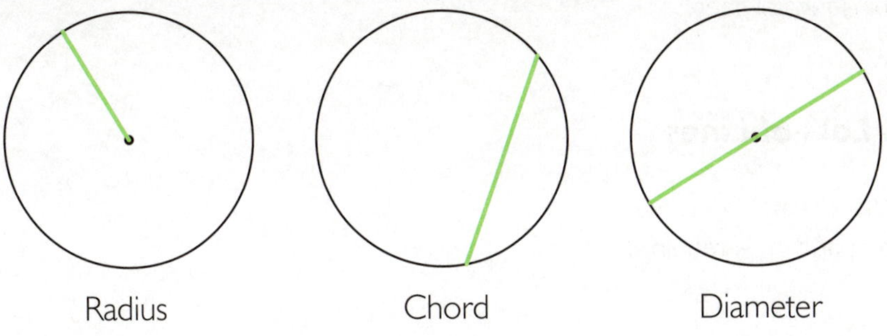

Radius   Chord   Diameter

▲ *A diameter is a special type of chord that is equal to two radii.*

12. Lines, Line Segments, and Rays

Preparation
No preparation is required.

■ **Materials**
- Ball of string
- Playdough

Activity

1. Lead a discussion about the mathematical meaning of terms such as "line". Say, *One way of thinking about lines is to imagine that a line is like a piece of string that goes on forever in both directions. It has no ends and stretches to infinity.* Invite a volunteer to hold the ball of string while someone else takes a free end and walks out of the classroom.

2. Say, *When we talk about a line segment, we are referring to part of a line. A line segment has end points.* Roll the playdough into two small balls and stick the balls on the string, about 30 cm apart. *Whenever we draw what we call a line, we are actually drawing a "line segment". We can imagine that the line segment could continue forever beyond its end points, like the string continues past the playdough. Most of the time though, we just ignore the rest of the line.*

3. Remove one of the balls of playdough and say, *A "ray" is similar to a line and similar to a line segment. It has an end point but continues onward forever.*

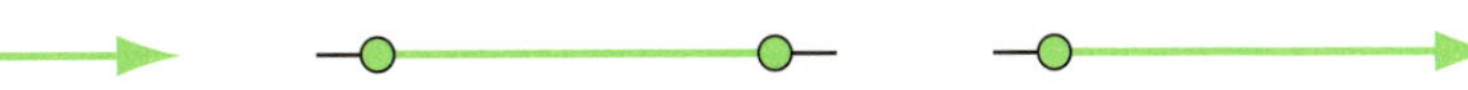

▲ *Discuss the mathematical concepts of lines, line segments, and rays with your students.*

combining lines

Combinations of lines can be seen everywhere. In the following activities students will make and investigate many interesting designs and puzzles made by particular combinations of lines.

■ Materials
- Toothpicks
- Tagboard or light card — 1 sheet for each student
- Glue

1. Lots of Lines

Preparation
No preparation is required.

Activity
Invite the students to explore how many "open" and "closed" shapes they can make from a given number of straight lines. Have them glue the shapes they like onto the tagboard for display. As an extension, have the students glue a design onto thicker tagboard to make stamps for creating embossed borders on sheets of paper.

Three lines   Four lines   Five lines

▲ *Students will discover the many and varied designs that can be made with just a few straight lines.*

■ Materials

None

2. Dot-to-Dot

Preparation
No preparation is required.

Activity
On the board, draw the dot-to-dot puzzles shown below. Ask the students to draw what they think the final shape for a particular puzzle will look like. Afterward, invite volunteers to physically join the dots to reveal the figure. Complete these puzzles with the students, then encourage them to make puzzles for each other.

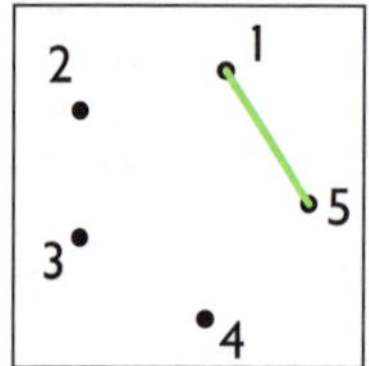 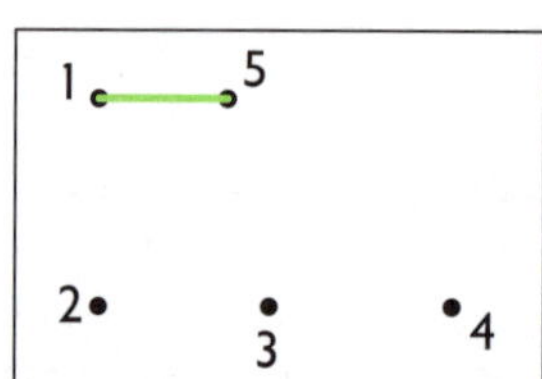 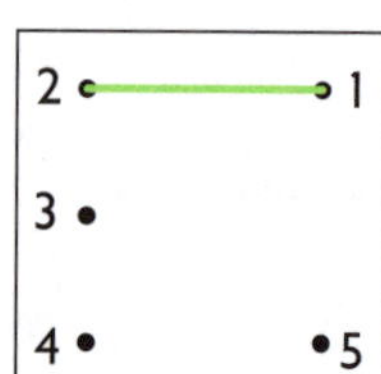

▲ *Ask students to imagine then sketch what a dot-to-dot puzzle will reveal to expand their visual thinking skills.*

3. Toothpick Puzzles

Preparation

Make an overhead transparency of Blackline Master 4.

Activity

Toothpick puzzles are a fun and challenging way to develop understanding of how shapes may be altered by moving or removing lines. Display the puzzles on the overhead transparency and invite the students to solve them.

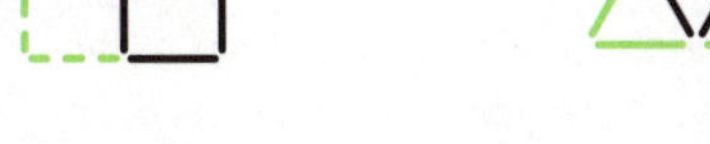

Make 2 squares

Make 4 small triangles and 1 large triangle

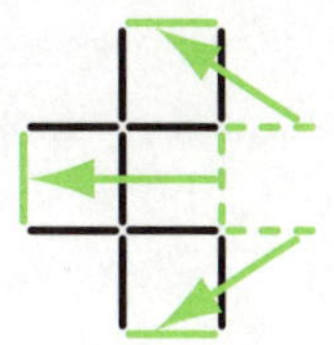

Make 3 squares

▲ *The above diagrams show the solutions to the toothpick puzzles on Blackline Master 4.*

■ Materials

- Blackline Master 4 (page 62)
- Overhead projector and blank transparency sheet
- Toothpicks

4. Plaid Patterns

Preparation

No preparation is required.

Activity

1. Show the students the plaid or checkered patterns. Ask, *What do you notice about the lines?* Bring out the fact that individual lines are parallel and repeating, some are horizontal and some are vertical, and (depending on the pattern) some are thicker than others.

2. Invite the students to design some plaid patterns. The patterns may involve horizontal and vertical lines, and the students may use the edges of the paper as a guide. Point out that, as most rulers have parallel sides, an easy way to draw parallel lines is to align one side of a ruler with an existing line, and then draw a line along the other side of the ruler.

3. Some students may like to make patterns with oblique lines. Encourage the students to explore different ways of using color and lines of different thickness in their designs.

■ Materials

- Plaid or checkered patterns

Did You Know?

A straight line extends in either direction for an infinite distance into space. A line segment is part of a straight line and has two end points. In everyday use, it is usually called a "line".

▲ *With simple combinations of parallel lines, the students can make attractive plaid patterns.*

■ Materials

- Blackline Master 5 (page 63)

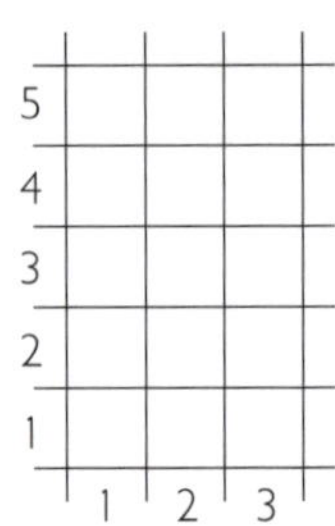

5. Block Writing

Preparation

Make a copy of Blackline Master 5 for each student.

Activity

1. Students may be familiar with "bubble" or "block" writing. Placing block writing on a grid provides the students with a format with which to extend the possibilities. A basic grid pattern is five units high and three units wide.

2. Altering distances between horizontal and vertical line segments on the grid produces interesting results, as does using curved line segments. Draw some of the examples shown below and invite the students to copy them or try others. Challenge adept students to make an entire alphabet on a grid.

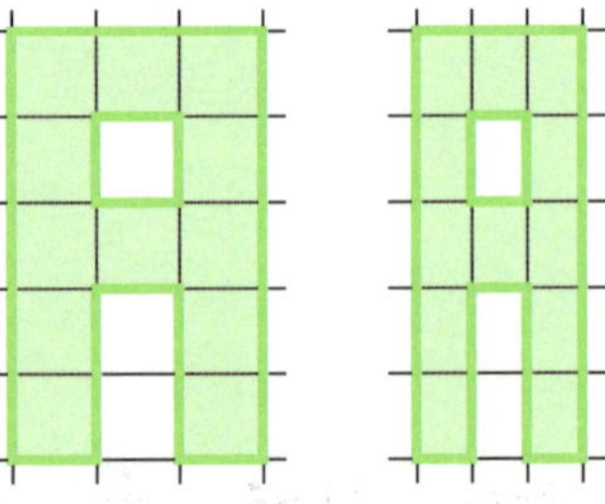 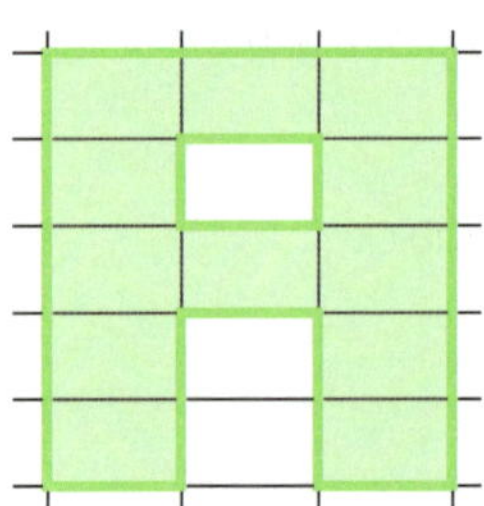 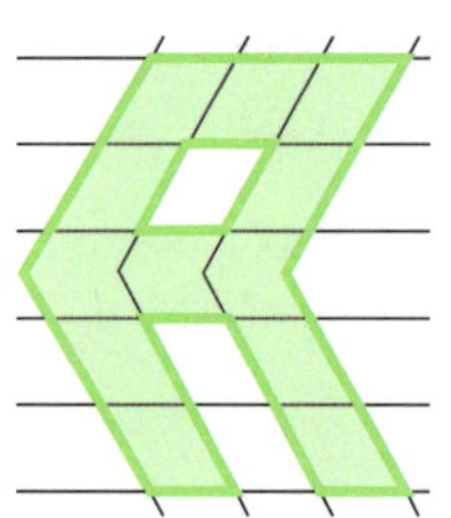  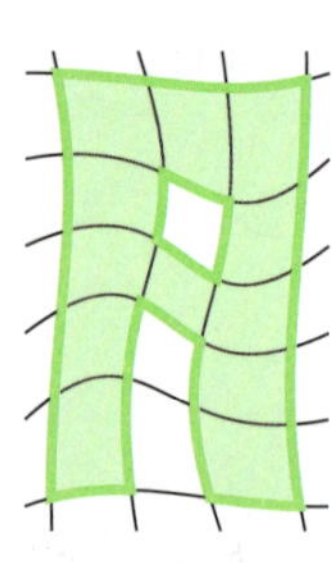

▲ *By changing the properties of a grid, the students can draw many varied and intriguing letters.*

10–12   

■ Materials

- Blackline Master 6 (page 64)
- Overhead projector and blank transparency sheet

geo

Pictures of 3D shapes display certain techniques to convey the illusion of depth. See *Paper Polygons* and *Faces and Frames* for activities on creating 2D pictures of 3D shapes.

6. Illusions

Preparation

1. Make an overhead transparency of Blackline Master 6.

2. Make a copy of Blackline Master 6 for every student.

Activity

1. Show the students the illusions from Blackline Master 6 on the overhead projector. Display the illusions one at a time and discuss the students' reactions.

2. Distribute the copies of Blackline Master 6 to the students and instruct them to use a ruler to help answer the questions.

3. Invite each student to try recreating the Poggendorf illusion. Display the results so that the students can compare the effectiveness of their copies. Investigate the illusion further by considering the following questions:
   - *If the three line segments are near-horizontal, what effect does that have?*
   - *Does the illusion work better if the vertical parallel lines are close or far apart?*
   - *Does the illusion still work if the vertical parallel lines are changed so that they are horizontal?*
   - *What happens if the vertical parallel lines are different lengths?*

The other illusions can be recreated and investigated in a similar way.

7. Chords Designs

Preparation

Make two copies of Blackline Master 7 for each student.

Activity

1. Have each student join dots on the sheet to sketch a circle and draw a diameter. The students can then pick a point on the circumference of the circle and draw chords to join it to each end of the diameter.

2. Ask, *What type of angle do you see where the two chords meet each other?* (A quarter angle.) Direct the students to repeat the procedure with a different point on the circle. Ask, *What do you notice about the angle formed this time?* (It is also a quarter angle.)

3. As the students continue the process of drawing chords, they will notice that a quarter angle is always formed when chords starting at each end of a diameter meet at a common point on a circle's circumference. Invite the students to shade the shapes within their diagrams to create dazzling designs.

Materials

- Blackline Master 7 (page 65)

Did You Know?

A circumference is the boundary of a circle.

A diameter is any straight line segment that passes through the center of a circle and joins two points on the circumference. A diameter is equal to two radii.

A chord is a line segment that joins two points on the circumference of a circle.

A quarter angle is one-quarter of a complete turn around a point. It is equal to 90°.

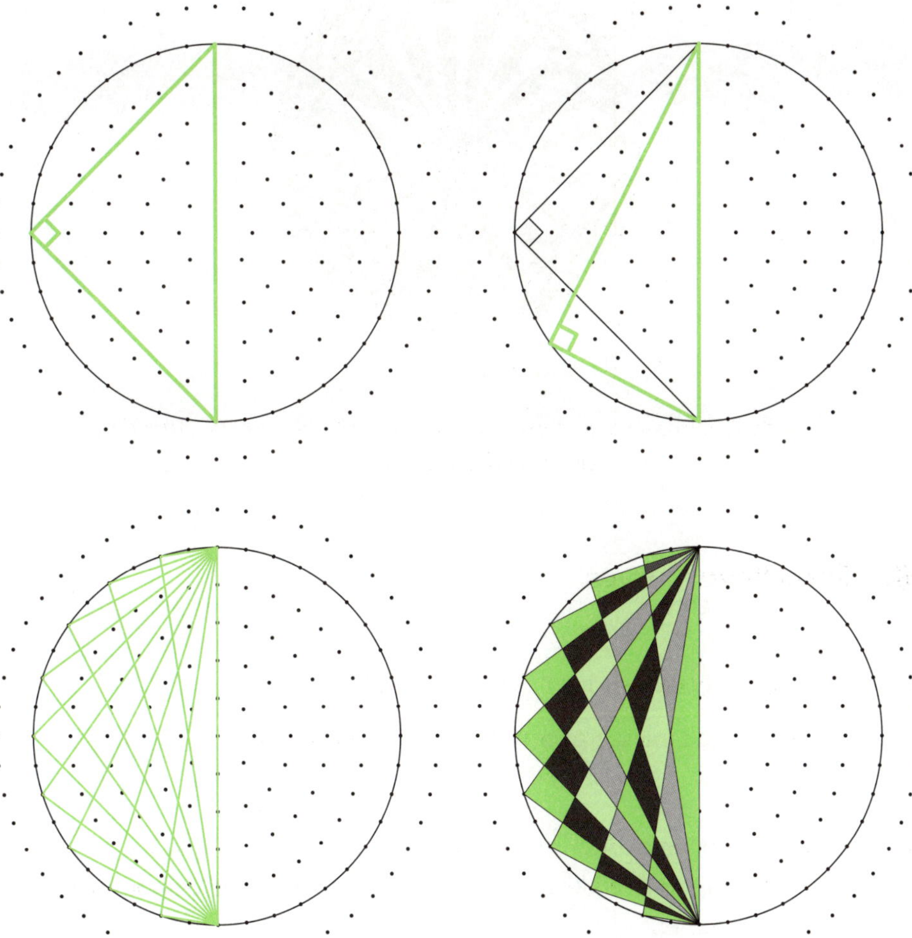

▲ *The students will discover an interesting property of certain chords, while creating stunning designs.*

geo Changing some properties of a shape without altering others is an aspect of topology. Dozens of fascinating activities on topology can be found in *Twists and Turns*.

geo

For many more fascinating activities on circles, see *Plane Puzzles*.

4. As an extension to the activity, have the students follow similar steps to those above, but with a quadrant instead of a semicircle.

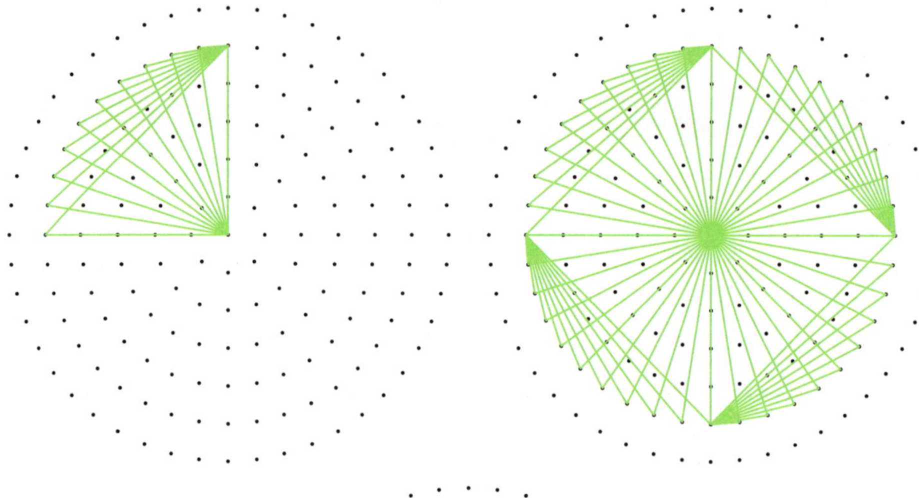

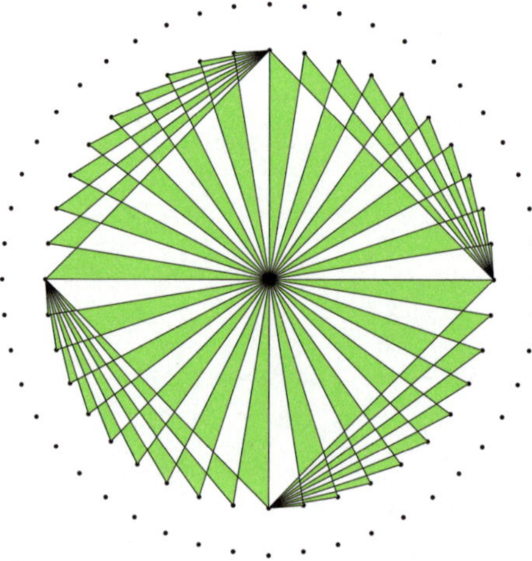

▲ *By following similar steps to those used for a semicircle, students create dazzling designs using quadrants.*

10–12

Materials
- Blackline Master 8 (page 66)
- Overhead projector and blank transparency sheet

8. Snowflake Curves

Preparation
1. Make a copy of Blackline Master 8 for each student.

2. Make an overhead transparency (OHT) of Blackline Master 8.

Activity
1. Distribute the copies of Blackline Master 8 to the students. Display the OHT of Blackline Master 8 in landscape format on the projector and demonstrate the following procedure:

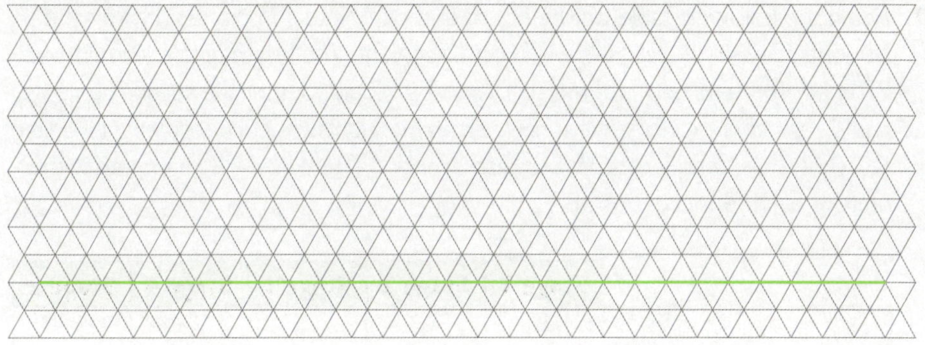

a. Near the bottom edge of the paper, draw a line that is 27 units long.

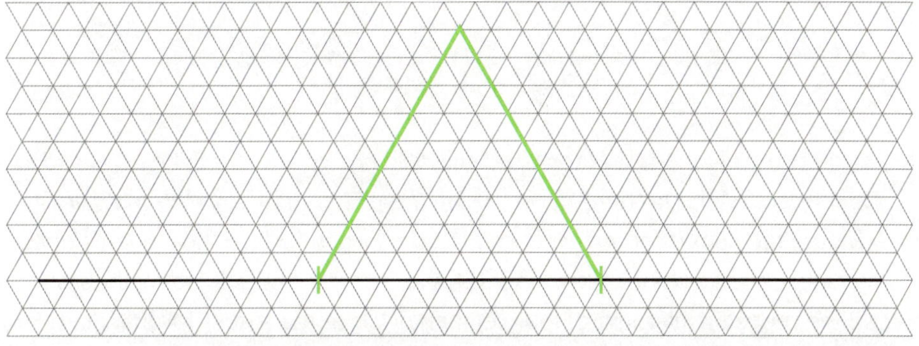

b. Divide the line into 3 equal segments (9 units each). Use the middle segment as a base and draw two extra line segments to form an equilateral triangle.

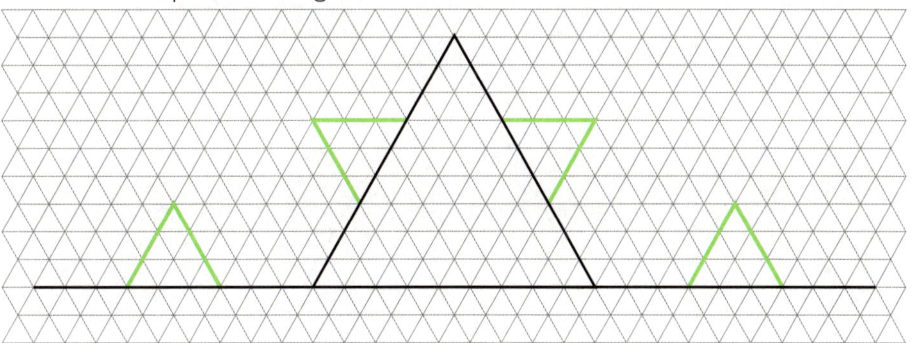

c. Divide each of the line segments (excluding the triangle base) into 3 equal segments. Add line segments to form equilateral triangles on each middle segment.

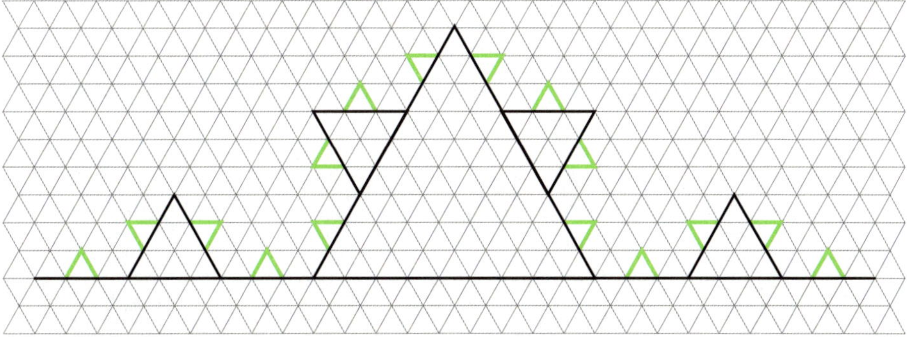

d. Repeat the previous step.

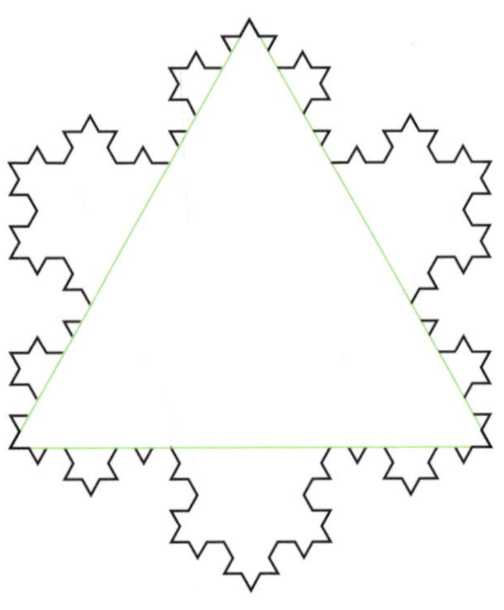

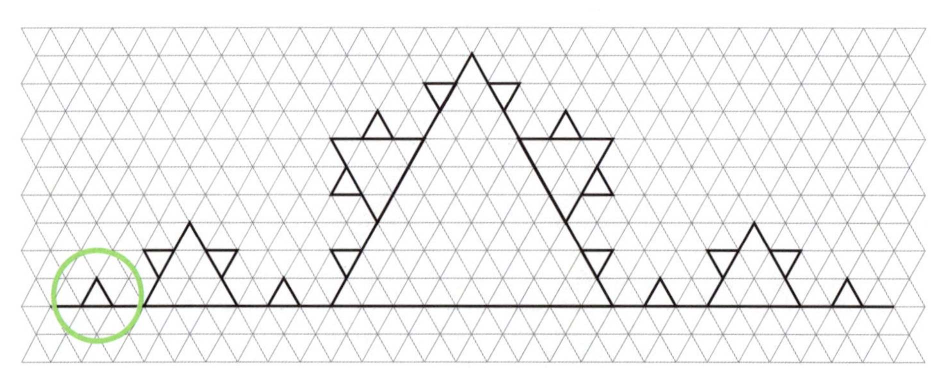

For other activities involving
mathematically similar shapes,
see *Paper Polygons, Plane
Puzzles,* and *Creative Coverings.*

2. Draw the following table on the board and complete it in discussion with the
students. Say, *In the first stage we had 1 segment. The length of the segment was
27 units. The total length of the segment was 27 units. After the second stage we
had 4 segments excluding the triangle base. How long was each segment?* (9 units.)
So, what is the total length of all the segments? (There were 4 segments with 9
units in each. 4 lots of 9 is 36.)

	Stage 1	Stage 2	Stage 3	Stage 4	Stage 5
Number of segments	1	4	16	64	
Length of each segment	27	9	3	1	
Total length	27	36	48	64	

▲ *The students can use a table like this to investigate
the figure they are drawing.*

3. As you complete the rest of the table, say, *It's not very fast to count each individual
line segment. What patterns do you see in the picture that can help you quickly
calculate the number of segments and the total length?* Have the students discuss
their ideas in pairs before conferring with the rest of the class. One method is to
use the repeating element of four segments. Once the results for that element
are calculated, the answer can be multiplied by the number of times the
element is repeated.

▲ *By counting the number of times the circled element repeats,
the students can more easily calculate the total number
of segments in the overall shape.*

4. After the numbers for Stage 4 are figured out, ask, *What patterns can you see
in the numbers?* (The number of segments in each stage is four times larger than
in the previous stage. The length of each segment is a third of what it was in the
previous stage.) *What do you think the answers for Stage 5 will be?* (256 line segments,
each segment will be 0.3 of a unit long, and the total length will be 85.3 units.)

introducing angles

This unit provides a sequence of lessons to introduce the concept of angles to students of all ages. It is recommended for students who are unfamiliar with angle concepts or for those who need a review. The basic knowledge that these activities develop will assist the students make the most out of the other angle activities in the book. These activities explore angles between two lines that are fixed (static) or changeable (dynamic).

1. Corners

Preparation
Make an oblong using the pipe cleaners and straws.

Activity

1. Hold up the oblong and a square pattern block. Ask, *What is the same about these two shapes? How are they different?* Bring out the fact that both have four sides but are different shapes and sizes. Identify a corner in each shape and ask, *What do you notice about the corners of both of these shapes?* (They look the same.) *How can you prove it?* (By overlapping them so that the two sides of each corner match up.)

2. Say, *There is something special about the way the sides join together that makes these shapes look the way they do. Each corner has two sides. The sides join together at a point and there is a certain amount of "opening" between the sides that gives the corner its shape.*

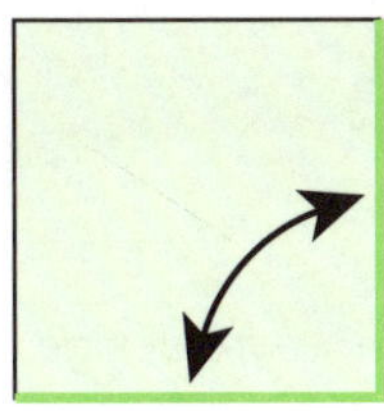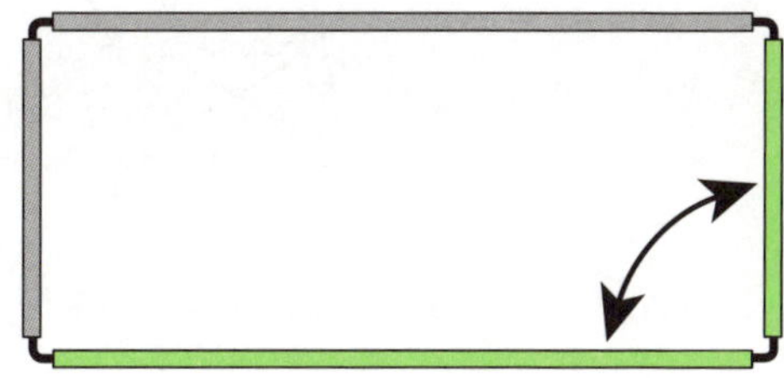

▲ *Invite the students to compare a corner of a square pattern block with a corner of an oblong made from drinking straws.*

3. Have each student find a shape in the classroom that has a corner with two straight sides that join together at a point. Afterward, invite volunteers to show the shapes they found and ask them to identify the corner they looked at, the sides of the corner, the point where the sides meet, and the general region between the lines.

■ Materials
- 4 pipe cleaners
- 4 drinking straws (2 pairs, each of different lengths)
- Pattern blocks

Did You Know?
An oblong is a rectangle with adjacent sides of different lengths. It is also known as a "non-square" rectangle.

Did You Know?
The word "angle" comes from a Latin word, *angulus*, which means "a little bending".

■ Materials

None

Did You Know?

A polygon is any simple two-dimensional closed shape formed by three or more straight line segments.

2. Fixed Angles

Preparation

No preparation is required.

Activity

1. Say, *Each corner in a pattern block has two straight sides that make the corner. If we draw a pattern block we just use straight lines to make the corners. We can find other places where two straight lines join together to make a corner.*

2. Have the students find examples in the classroom, such as desks and books. Afterward, take the students for a walk around the school grounds. Challenge them to find other examples of two straight lines coming together at a point. Examples can be found in play equipment and buildings, in the twigs of trees, in letters of the alphabet, or on road intersections shown on street maps. Try to avoid using rounded corners as examples; that is, where there is no clear joining point between the sides.

3. Introduce the terms "angle arms" and "vertex" to describe the "lines" and "joining point" respectively.

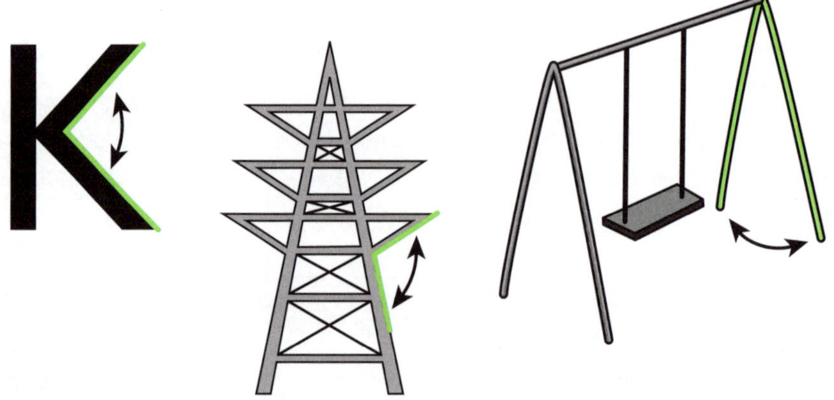

▲ *Encourage the students to identify places where they can see fixed angle arms.*

■ Materials

- Thick pipe cleaners — 1 for each student
- Drinking straws — 1 for each student

3. Comparing Fixed Angles

Preparation

No preparation is required.

Activity

1. Say, *If you want to compare the shape of a corner or the opening of one shape with another, how can you do it?* Discuss the students' ideas and allow them to try their ideas by using any materials available.

2. The simplest method for comparison that the students may suggest, is to overlap two corners. Ask a student to demonstrate and explain the process that he or she uses. If necessary, highlight the fact that the vertices of each corner must be aligned. Also, one angle arm of a corner must be aligned with an angle arm of the other corner.

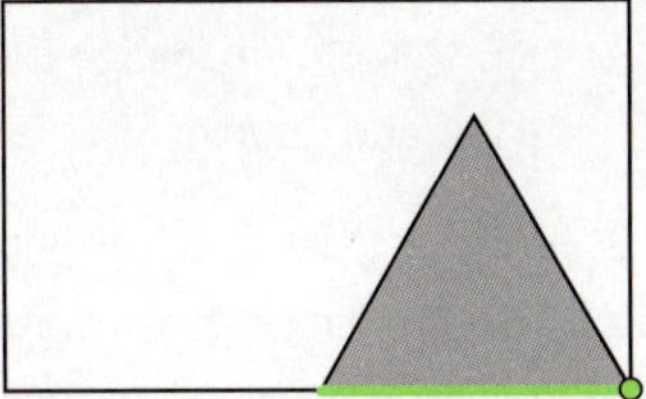

▲ *One method that the students can use to compare angles is to overlap the angle arms.*

3. Another method the students may suggest is to draw around the angle arms that form a particular corner. In this way, the outline of the corner is recorded so that the angle arms are represented by distinct lines and can be clearly identified, along with the vertex. Have the students use this method to record several corners. They can then compare each new corner they investigate with the ones they have drawn.

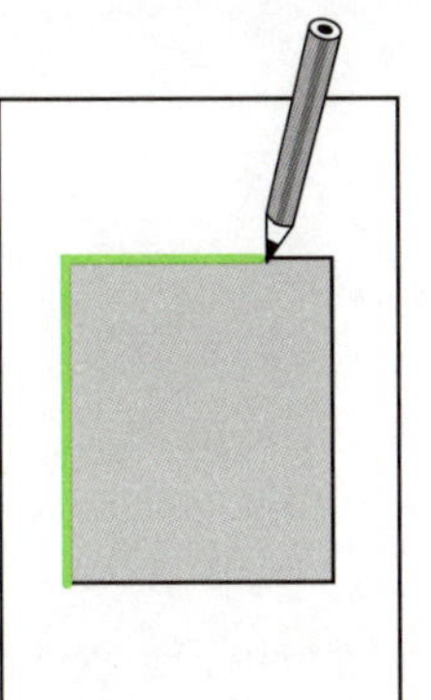
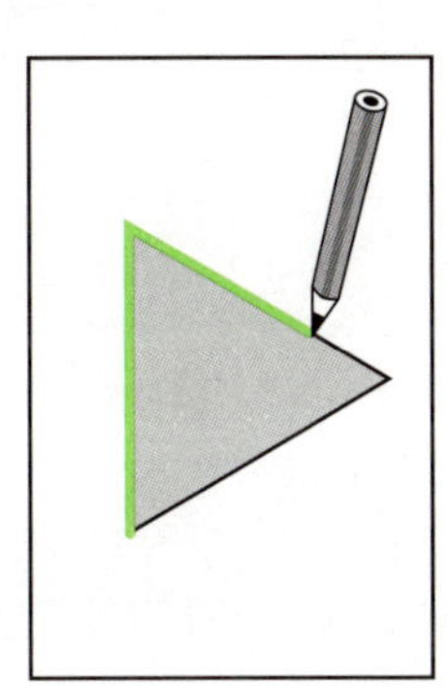

▲ *Drawing around the two angle arms that make a corner is another way students can directly compare angles.*

4. One technique the students may not suggest is to use a tool of some kind that can be carried from one corner to another. Hold up the straw and push the pipe cleaner inside it. When the straw is bent over, the pipe cleaner should make it hold its shape. Have the students make their own "portable corners", which they can then fit against a range of objects in the classroom.

5. As individual students compare corners, ask them to describe whether the opening between the angle arms in one is wider than in the other. Ask, *Does the length of the angle arms affect the size of the opening?* Have the students provide examples to prove their opinion. If necessary, compare the corners of a square pattern block and a large book to demonstrate how the length of the angle arms does not affect the size of the opening.

Did You Know?
A "vertex" is the point where two or more line segments join or intersect on a 2D shape, where three or more edges meet on a 3D shape, or the point where two or more angle arms intersect.

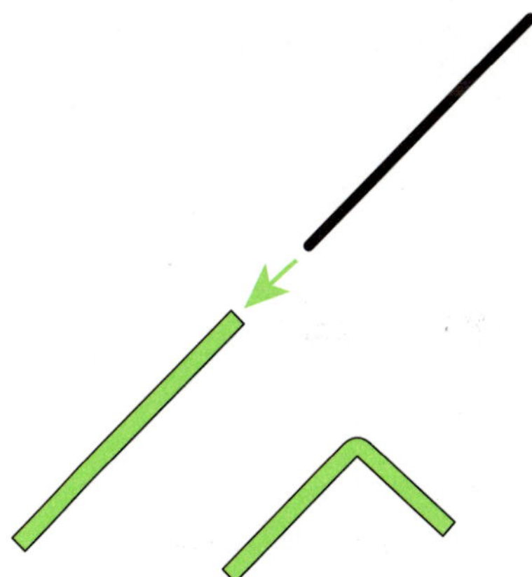

▲ *The students can make a "portable corner" to record the amount of opening between two angle arms.*

Materials

- Large book

4. Representing Fixed Angles

Preparation

No preparation is required.

Activity

1. Have the students choose three shapes. Challenge them to develop a way of representing the angle arms, vertex, and opening of a corner of each shape on paper.

2. Discuss the students' ideas. It is likely that they will find representing the angle arms and vertex easy to accomplish, either by drawing the outline of the corner or using a similar method. Representing the opening between the arms may be more difficult. Some students may shade the area between the arms, while others may be unable to think of a way to represent the opening.

3. Place a large book on the board and draw around the outside of one of its corners. Point out the angle arms and use a different color of chalk to mark the vertex. Explain that the opening between the angle arms can be represented by using a double-headed arrow. Complete the diagram as shown below.

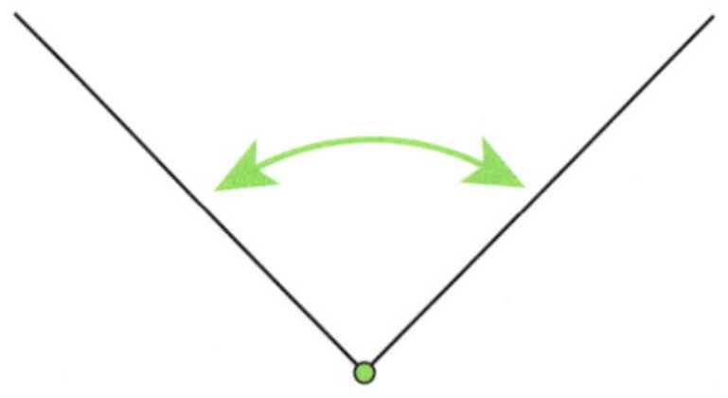

▲ *Once students have explored real-life examples of fixed angle arms, they will be more ready to draw representations such as this one.*

4. The students can practice drawing around and marking other corners in a range of orientations.

Materials

- Pattern blocks
- Geostrips and fasteners or similar

5. Measuring Fixed Angles

Preparation

No preparation is required.

Activity

1. Compare each corner in a single type of pattern block with the others in that same type of pattern block by overlapping or drawing around the shapes. The students will find that the corners are all the same shape in the triangle, square, and hexagon pattern blocks. They will also discover that the blue rhombus has two different corners, as does the pale rhombus and red trapezoid.

▲ *Have the students investigate the corners of pattern blocks.*

2. Hold up a blue rhombus and ask, *What can we call the different types of corners in this rhombus?* Discuss the students' ideas and suggest using "wide corner" and "narrow corner". Link these terms to the relevant corners in the pale rhombus and the red trapezoid.

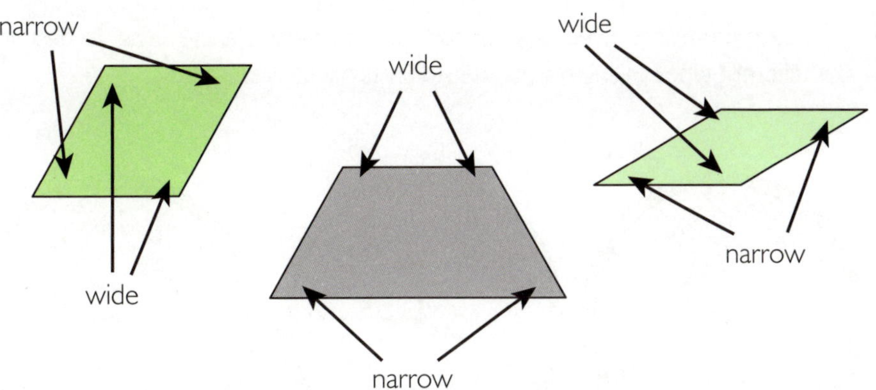

▲ *An informal way of describing the different corners in these pattern blocks is to use the terms "wide" and "narrow".*

3. Have the students find five corners of objects to measure. (They may like to make some unusual 2D shapes using the geostrips, as classrooms tend to have a lot of objects with right angles only.) The students should use the pattern block corners as tools to measure the corners of their geostrips shapes. As the students compare the pattern block corners with the other objects, ask individuals questions such as, *What corners are the same as the corner of a square pattern block? Can you find a corner that is like the wide corner in the blue rhombus? How many narrow corners of the pale rhombus fit into the corner of a book cover?*

4. Afterward, ask, *Which pattern block corner did you find the most examples of? The least? If you could only use one pattern block corner to measure all corners which one would you use? Why?*

Did You Know?
In North America, the term "trapezoid" is used to describe a quadrilateral that has only two parallel sides. In Australia, Europe, and elsewhere, the same shape is called a "trapezium".

■ Materials

None

6. Changeable Angles

Preparation

No preparation is required.

Activity

1. Say, *Some objects have angle arms that move to show different types of openings or corners. What are some that you can see in the room?* Discuss the students' suggestions, pointing out, if necessary, examples like scissors, clock hands, and the students' own bodies (for example, fingers or an arm moving away from the side of the body).

2. Challenge the students to identify the vertex in each instance they find. It can be difficult to identify the vertex exactly with objects that have thicker angle arms. In such instances, have the students identify the region where the vertex would be.

3. Direct the students to investigate the amount of turn each object can make from where the two arms are at the same starting position to where the opening between them is as large as possible. Discuss the results as a class; have the students identify the opening they are talking about.

4. Introduce the terms "fixed angle arms" and "changeable angle arms" to describe the different types of angle arms the students have investigated.

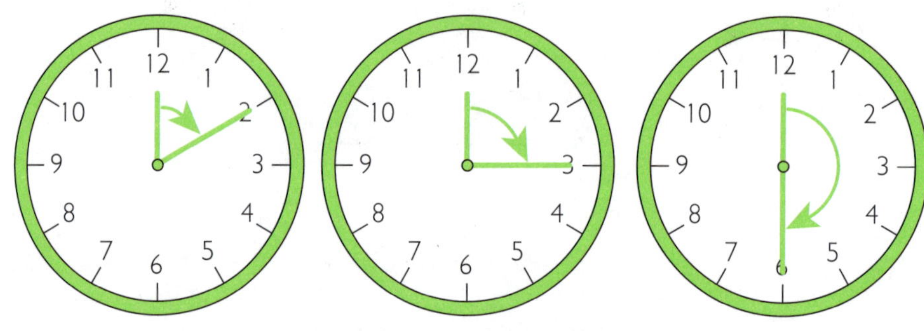

▲ Clock hands are an example of changeable angle arms.

■ Materials

- 2 geostrips with fastener — 1 set for each student and 1 set for demonstration

7. Representing Changeable Angles

Preparation

Join the geostrips together at one end.

Activity

1. Show the students that there are two geostrips. Arrange the strips so that one lies exactly over the other (shown at right).

Say, *These two strips are starting in the same position. I can turn one of them a certain amount of turn.* (Rotate one strip a quarter turn as shown at right.) *How can we draw a picture to show how much this strip turned from the other one?* Have the students discuss their ideas with a partner and then have them each draw a picture to represent the amount of turn.

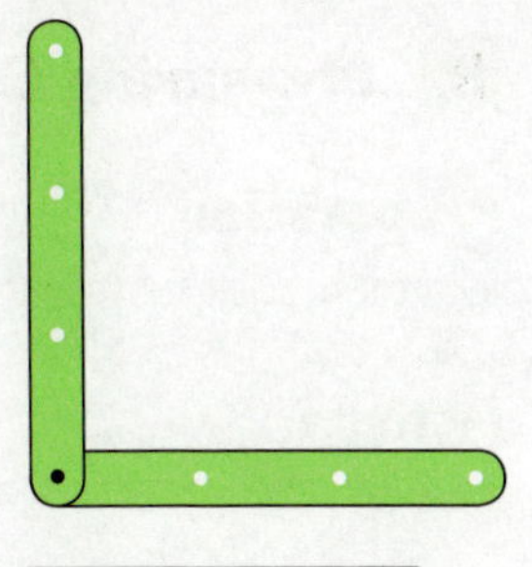

Did you know?
Throughout history, many different symbols have been used to represent angles. They include the following:

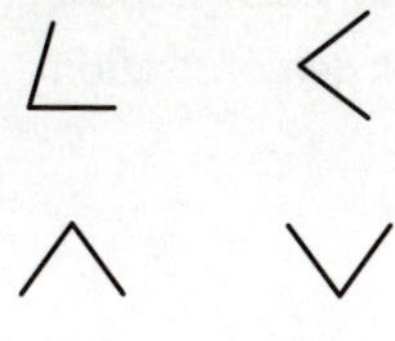

2. Discuss the students' pictures and say, *One way we can show how much the strip moved is to first show where it started with the other one. Then we can draw where it finished. We can also mark where the vertex is: it's where the fastener is on the geostrip.*

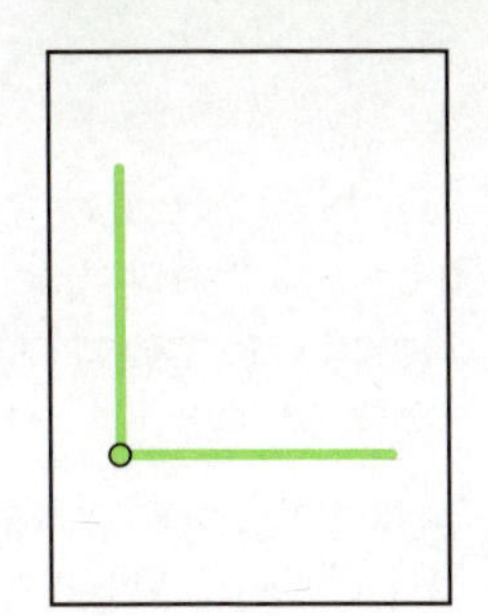

3. Say, *We can see that there's a certain opening between the two strips. How can you show the opening between two angle arms?* (By using a double-headed arrow.) Ask the students to add an arrow on their diagram and point out that the diagram is looking very similar to the ones used to draw fixed angle arms.

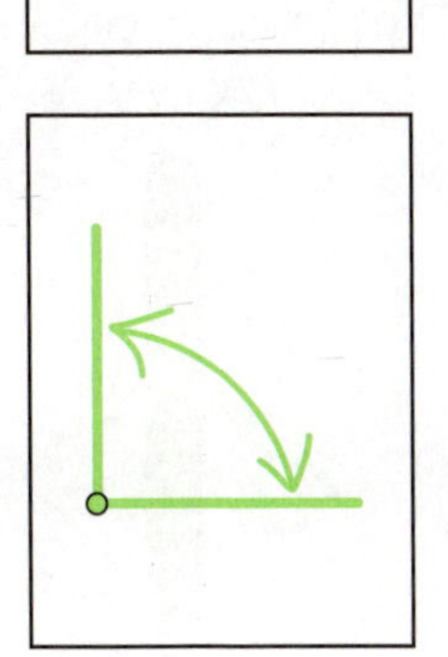

4. Say, *We can use the arrow a different way. The arrow can also show in which direction the angle arm moved.* Draw your diagram again so that only a single arrowhead is used.

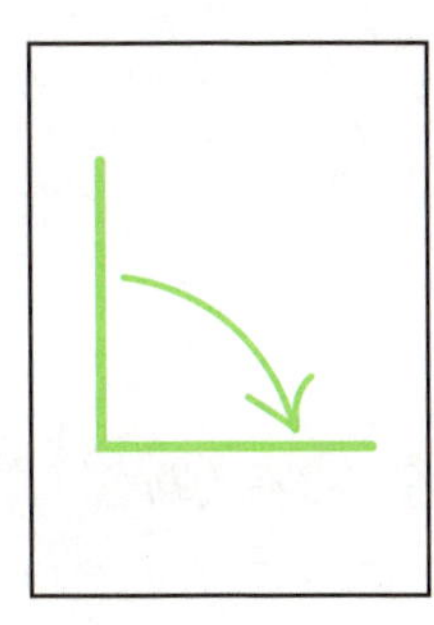

More **activities** on rotation can be found in *Making Moves* and *Simple Symmetry*.

5. Realign the strips then move one strip to a position different to that shown previously. Direct the students to draw a diagram showing the opening, or amount of turn. Ask, *How can we be sure to draw the correct position of the two geostrips in the picture?* Bring out the idea of drawing along the inside of the strips, as was done previously for representing fixed angles. Have the students use two geostrips to make their own moveable angle arms, align the arms over one another, then move one arm to a new position. They can then illustrate the arrangement by drawing alongside the inside edge of the strips.

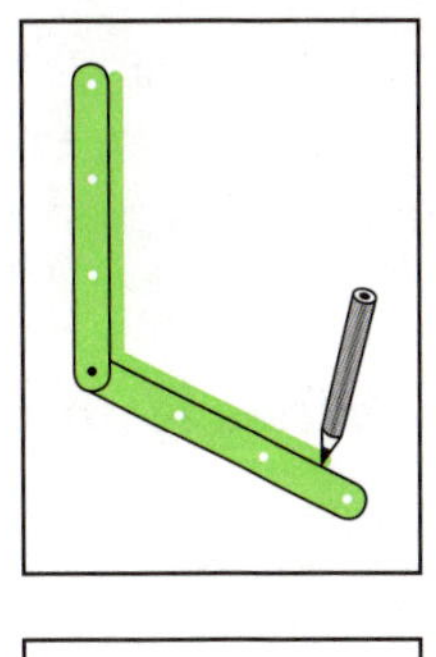

Materials

- 2 geostrips with fastener — 1 set for each student and 1 set for demonstration

8. Measuring Changeable Angles

Preparation

Join the geostrips together at one end.

Activity

1. Align the geostrips over one another. Hold one strip and rotate the other. Ask, *How can we measure the amount of turn that this angle arm made from where it started with the other?* Discuss the students' ideas.

2. Ask, *How do you describe the amount of turn that the strip makes if I do this?* Align the strips then rotate one strip completely around so that it ends up once again aligned with the other strip. Invite volunteers to describe their answers and introduce the term "one full turn". Have the students use their geostrips to make one full turn.

3. Repeat the previous step to introduce the concepts of "half a full turn" and "quarter of a full turn". When you feel the students are comfortable with these ideas, shorten the names to "half turn" and "quarter turn".

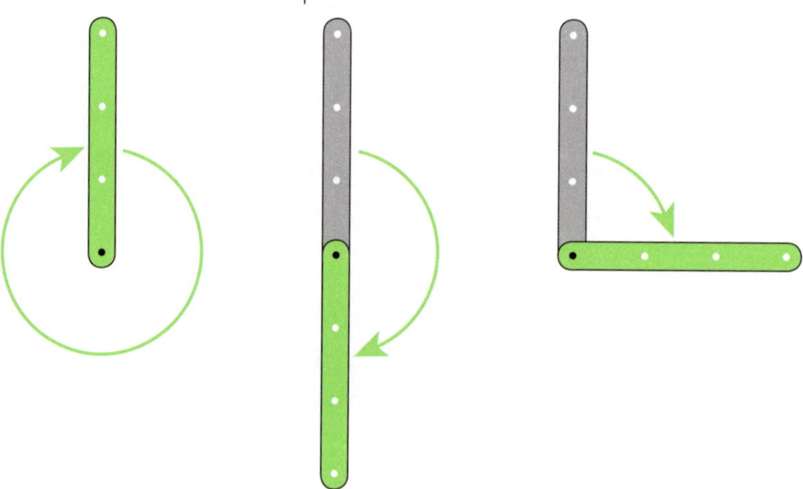

▲ *Use geostrips to introduce the concepts of a full turn, half turn, and quarter turn.*

4. Invite the students to explore objects in the room that have changeable angle arms. Challenge them to find angle arms that can perform a full, half, or quarter turn. They can also describe angle arms that can make amounts of turn between these reference amounts. For example, a pair of scissors may be able to open more than a quarter turn but less than a half turn.

Materials

None

9. What is an Angle?

Preparation

No preparation is required.

Activity

Encourage the students to consider their own definition for what an angle is or involves. The students may like to work in pairs to discuss their ideas. Do not expect them to write (or learn) a concise and complete definition, as the concept is quite involved. Take the time to discuss any erroneous ideas the students have, such as angle being related to area, or that larger angles have longer angle arms than smaller angles. For your reference, see the Glossary for a definition of angle.

geo measuring angles

Students often develop limited understanding of angle measurement until the protractor is introduced. Research suggests that, as with investigating other types of measurement, there should be a gradual progression from using non-standard units of measure (e.g. pattern block corners) to standard ones (degrees). The activities in this unit provide a sequence to develop a greater conceptual understanding of angle measurement. As such, students will gain the most benefit from completing the activities in the order that they appear. As the students explore new tools for measuring angles, provide opportunities for them to measure angles in real-life contexts. When you feel the students are ready, introduce the arc as a substitute for an arrow to mark the opening between angle arms.

1. Inside and Outside Angles

Preparation
Insert the pipe cleaner into the drinking straw.

Activity

1. Draw an oblong on the board and say, *Look at each corner of this oblong. The sides of the oblong make a pair of angle arms and a vertex.* (Indicate these features in one particular corner.) *We can examine the angle inside this corner of the oblong. We can also look at the angle that is outside this corner. Both angles have the same angle arms and the same vertex.*

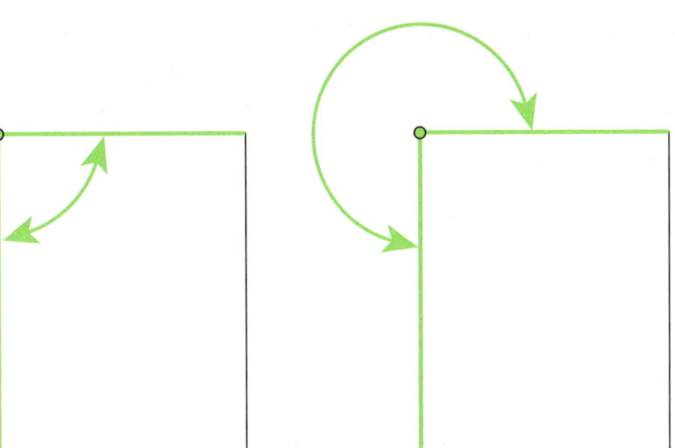

▲ *Highlight the fact that there are always two angles formed by a pair of angle arms. With polygons, there is an inside angle and an outside angle.*

Did You Know?
An oblong is a rectangle with adjacent sides of different lengths. It is also known as a "non-square" rectangle.

■ Materials
- Drinking straw
- Pipe cleaner

Did You Know?
"Exterior angle" can also refer to the angle supplementary to the interior angle. A "supplementary angle" is one that shares the vertex and one angle arm of another. The total of the angles is 180°.

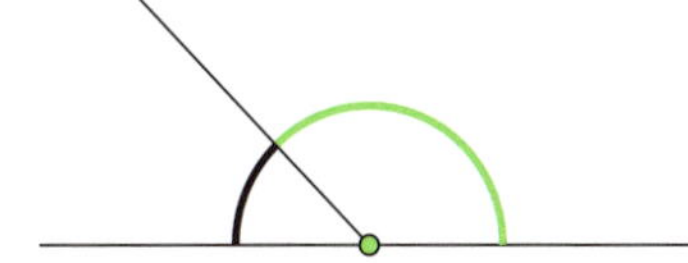

2. Ask the students to find examples of "inside" and "outside" corners in objects with fixed angle arms, for example books, desks, and pattern blocks. With older students, you may also introduce the terms "interior" and "exterior" to describe the types of angles.

3. Bend the drinking straw and say, *Some angle arms don't have an "inside" or "outside". All we can do is identify the angle we want to talk about. We show this on paper by marking the opening that we're interested in.* Draw a diagram on the board to demonstrate and have the students consider other instances.

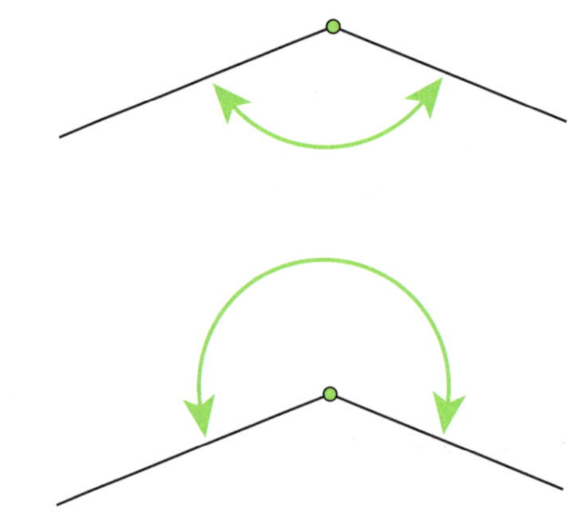

▲ *This diagram shows how we can hightlight different angles around the same vertex.*

■ Materials
- **2 geostrips with fastener**

2. Two Ways to Turn

Preparation
Join the geostrips together at one end.

Activity
1. Display the geostrips and identify the angle arms and vertex. Then align the geostrips over one another. Say, *There are two ways of turning one of these strips so that it ends up in the same position.* Move one of the strips a quarter turn while holding the other strip.

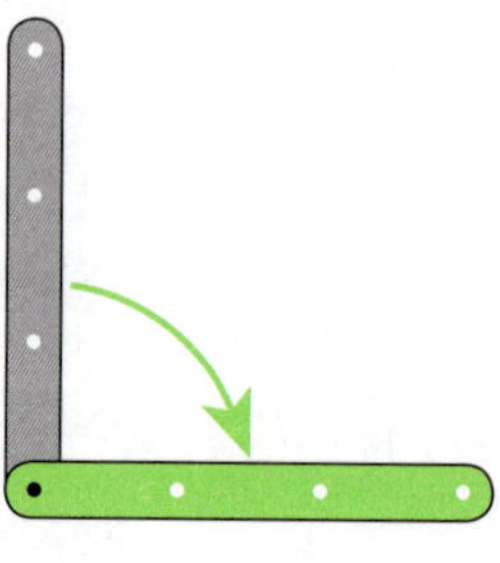

▲ *A quarter turn.*

2. Realign the strips over one another and ask, *If I turn the strip in the other direction, how much of a turn will I have to make so it finishes where it did before?* Move the strip as described and introduce the term "three-quarter turn" to describe the amount of turn.

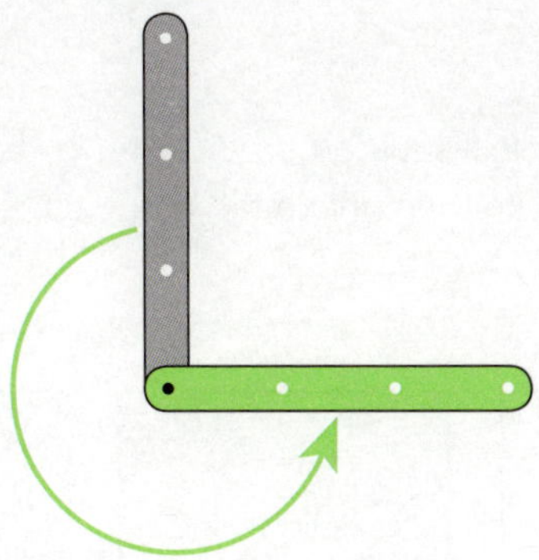

▲ *Discuss how the same finishing position can be achieved by turning in two different directions.*

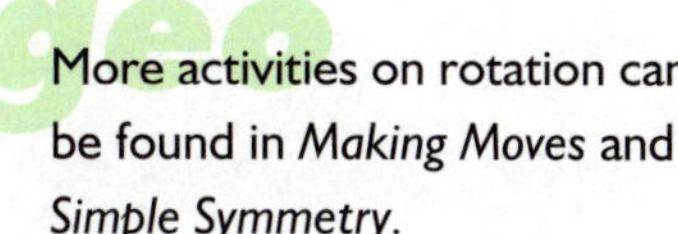

3. Discuss the language that can be used to describe the direction of turn, such as "left" and "right", or "clockwise" and "anticlockwise". Afterward, show the students how to represent the different turns on paper.

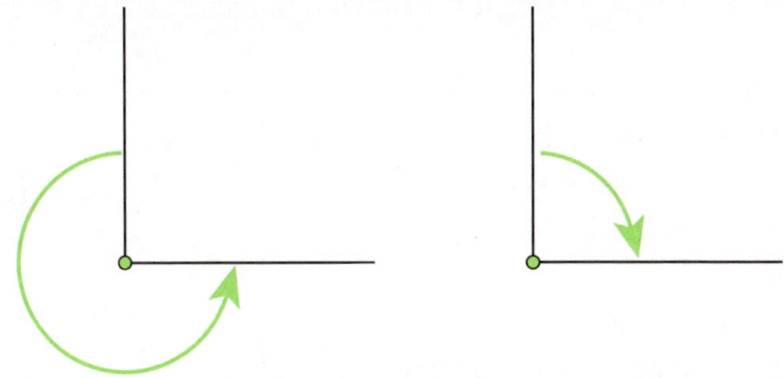

▲ *Show the students a method of drawing the two ways an angle arm can turn to finish in the same position.*

3. Combining Turns

Preparation
No preparation is required.

Materials
None

Activity
1. Have the students stand, facing the front of the room. Call out an amount of turn and have the students rotate that amount of turn to their left. Use quarter, one-half, three-quarter, and full turns. Repeat for several turns

2. Once the students are comfortable with these amounts of turn, start using simple combinations of turn. For example, say, *Turn two quarter turns* (that is, a half turn), *Turn two half turns* (a full turn), or, *Turn a half plus a quarter turn* (a three-quarter turn).

■ Materials

- Pattern blocks

4. Square-Corner Testers

Preparation

No preparation is required.

Activity

1. Have the students make a square-corner tester by folding a sheet of paper in half then half again. They can then unfold the paper to reveal the creases.

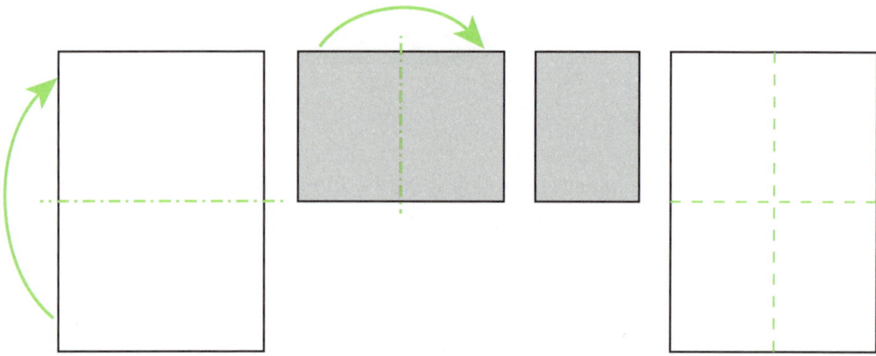

a. Fold the paper in half.

b. Fold the paper in half again.

c. Unfold the paper to show the creases.

2. Direct the students to draw a circle around the central intersection point where the creases meet on both sides of the paper. Folding the tester again, the students will find that they have marked a corner of the tester. They can then identify the angle arms and vertex of the marked corner.

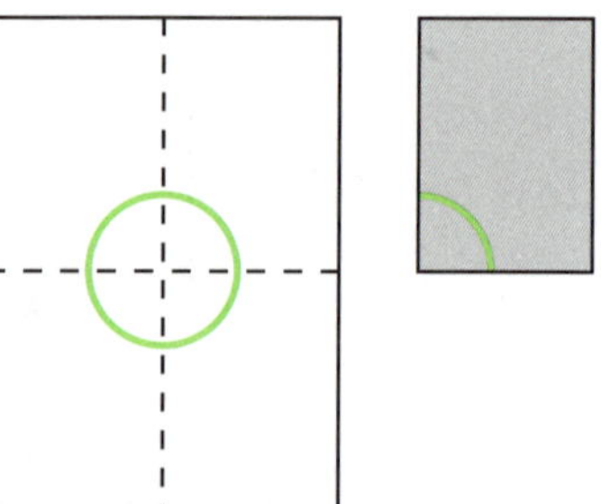

▲ *By seeing a fraction of a circle when they refold their paper, the students should start to see the link between a full turn and part of a full turn.*

3. Identify the angle arms and vertex of the marked corner and ask, *Which pattern block corners can fit exactly into the corner you marked?* Even though some students will know that the corner of a square pattern block will fit exactly, have the students investigate a variety of pattern block corners, including combinations of different blocks. In addition to the corner of the square block, three copies of the "narrow" (acute) corner of the pale rhombus can be arranged around the vertex without leaving a gap. Similarly, combining the corners of the pale rhombus and blue rhombus will also align.

4. Have the students find other examples of objects whose angle arms align exactly with the marked corner of the tester. They will find many examples in the classroom. Say, *All the corners you tested had the same amount of opening. What name could we give to this tester?* Introduce the term "square-corner tester" and relate the name to the corners of square pattern blocks.

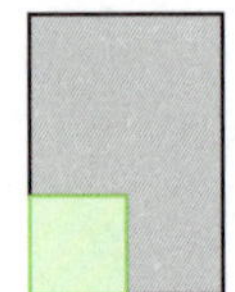

▲ *Have the students test which pattern block corners can be arranged between the angle arms of the tester.*

5. Quarter-Turn Tester

Preparation
Join the geostrips together at one end.

Activity

1. Align the geostrips over one another. Hold one geostrip and turn the other a quarter turn. Ask the students to identify the amount of turn.

2. Position a square-corner tester between the two edges of the geostrips as shown in the diagram below. Ask, *What do you notice?* (When you turn the strips a quarter turn, they have the same amount of opening as a square corner.) *So, what can we say about the angle arms in a square corner?* (They are a quarter turn apart.) *What's another name that we can give this tester?* (A quarter-turn tester.)

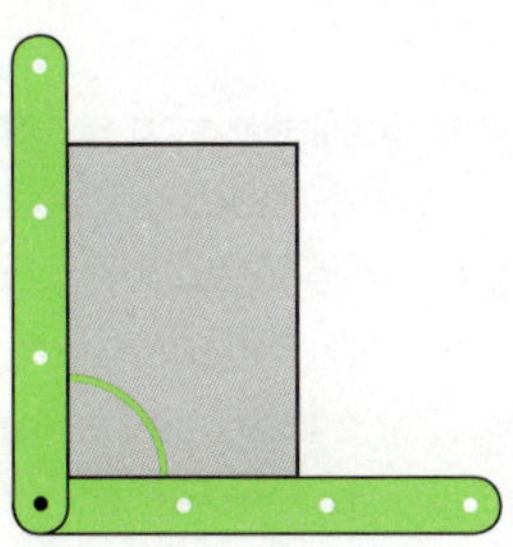

▲ *The students will see why a square-corner tester can also be called a quarter-turn tester.*

3. Ask, *How many quarter turns are there in one full turn?* (Four.) *How many quarter-turn testers do you think you can fit around a point?* (Four.) To reinforce the link between the amount of opening and the amount of turn, request testers from four students and use Blu-Tack to arrange them on the board. Align the geostrips over one another, then hold one of the geostrips and turn the other. As the moving geostrip passes the edge of each tester, have the students identify the amount of turn that the strip is making from its starting position.

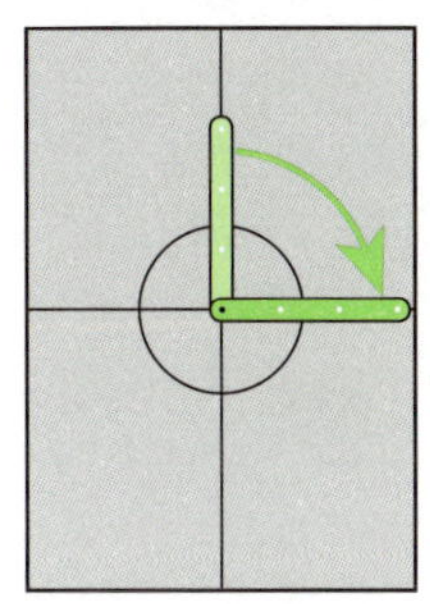 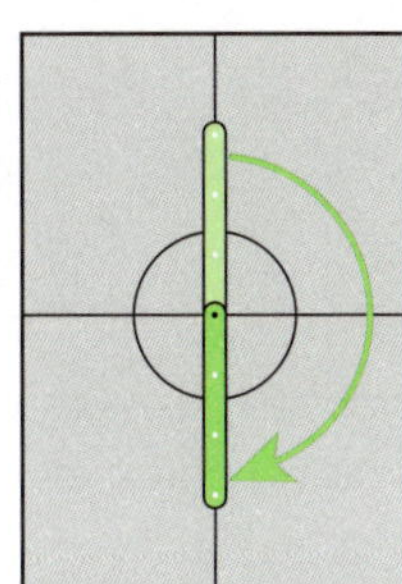 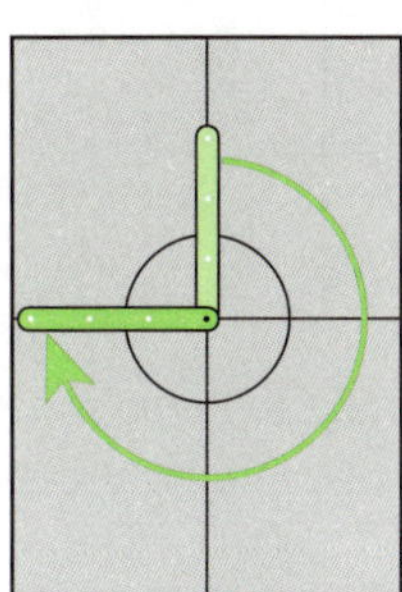 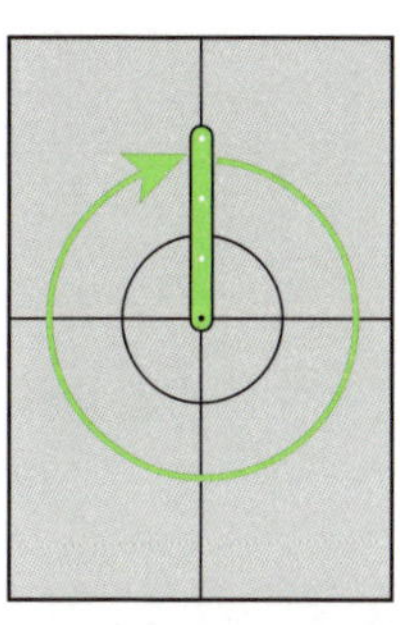

| 1 quarter | 2 quarters | 3 quarters | 4 quarters |

▲ *Using four testers provides the students with an opportunity to see how an amount of turn around a point can be measured using equal units.*

■ Materials
- 2 geostrips with fastener
- Square-corner testers — 1 for each student (reuse those from the previous activity)
- Blu-Tack

Did you know?
Throughout history, many different symbols have been used to represent angles that are a quarter of a full turn. They include the following:

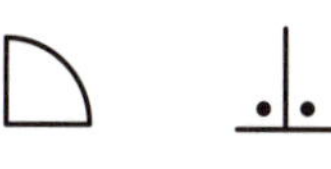

■ Materials

None

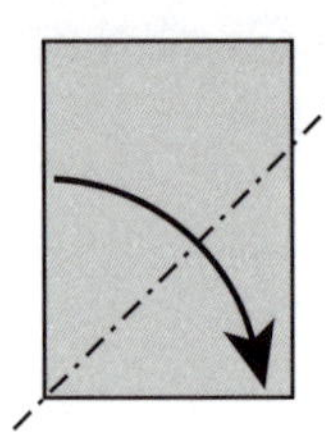

Fold a quarter-turn tester
to make these edges meet.

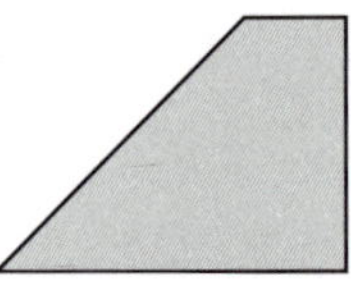

▲ *An eighth-angle tester can be
made by folding a quarter-angle
tester as shown above.*

6. Eighth-Angle Tester

Preparation

No preparation is required.

Activity

1. Direct the students to make a quarter-turn tester as described in Activity 5. Have the students identify the angle arms and vertex. Each student can fold his or her tester once more as shown left. Ask the students to identify the angle arms and vertex.

2. Have the students work in threes to investigate their new testers. Pose the following questions to guide them. Ask, *How many of these testers fit into a quarter-turn tester? How many testers fit around a point without leaving any gaps between the angle arms? What amount of a full turn do you think this tester shows? How do you know?* The students will discover that two of their new testers are equal to one quarter-turn tester.

3. To help demonstrate that the tester shows an eighth of a turn, ask the students to unfold the tester and count the number of corners that they can see made by the creases (there are eight). Have one student in each group refold their tester and examine the other members' unfolded testers to check that each corner has the same amount of opening. Then instruct the students to place a pencil on one of the creases and rotate it crease by crease, counting the number of times the pencil moves to return to the starting point.

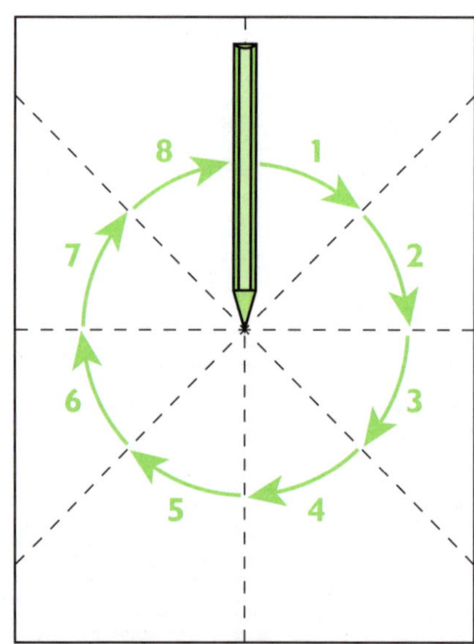

▲ *One method to demonstrate that a full turn around a point can
be divided into eight equal amounts of turn or openings.*

4. Say, *This tester shows eight equal corners around a point. One tester is an eighth of the total number of testers. The tester also measures one-eighth of a full turn around a point. Both amounts of turn and openings we see in corners can be described using the word "angle" so we call this tester an "eighth-angle tester". Similarly, the quarter-turn tester can be called a "quarter-angle tester".*

5. Have the student groups use their testers to draw examples of shapes that have a corner that is a one-eighth angle. Discuss the examples afterward.

Did you know?

One full turn around a point can be divided into any number of smaller, equal amounts. "Degree" is the word used to describe an amount that is 1/360 of a full turn.

 Multi-Angle Tester 1

Preparation

1. Make a copy of Blackline Master 9 for each student.

2. Make copies of Blackline Master 10 on color paper so that there is a tester for each student.

3. Make an enlarged copy of one tester from Blackline Master 10 for demonstration.

Activity

1. Direct the students to unfold their eighth-angle testers (from the previous activity) so that they become quarter-angle testers again. Ask, *How many eighth angles can we measure with this tester?* (Two.) *How can we make a tester that shows an angle that is four-eighths of a full turn? What about three-eighths?* The students will discover that when they unfold the tester again it shows four-eighths. Folding back one of the eighths will then show three-eighths. Ask, *Can you work out a way of showing five-eighths?* The students will find that they need to cut into their tester to show five-eighths.

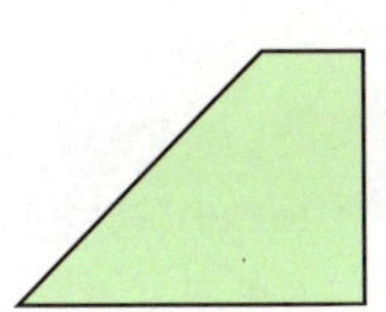 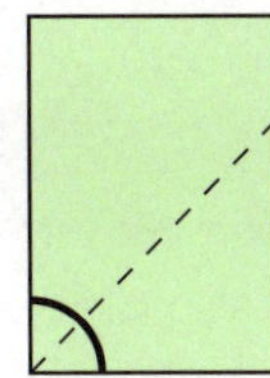 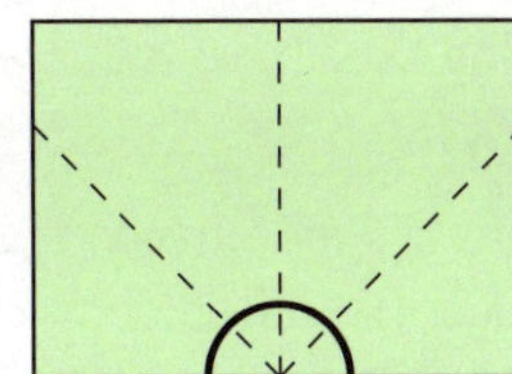 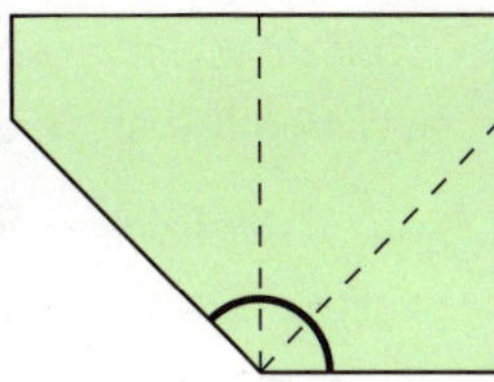

▲ *The students will see how an eighth-angle tester can be folded to measure one-eighth, two-eighths, four-eighths, and three-eighths of a full turn.*

2. Ask each student to cut out the eighth-angle tester and along its dotted line. By then folding the eighths under and over on themselves like a fan, the students will produce a tester that can show any combination of eighths. Ask the students to make sure that the tester lies as flat as possible.

Materials

- Blackline Master 9 (page 67)
- Blackline Master 10 (page 68)
- Students' eighth-angle testers from the previous activity
- Eighth-angle tester — 1 for demonstration
- Scissors — 1 pair for each student

Did You Know?

An interior angle is an angle inside the perimeter of a polygon. Its angle arms are two adjacent sides of the polygon.

A "vertex" is the point where two or more line segments join or intersect on a 2D shape, where three or more edges meet on a 3D shape, or the point where two or more angle arms intersect.

Angle arms are two line segments, real or imaginary, that share a common vertex and create an angle.

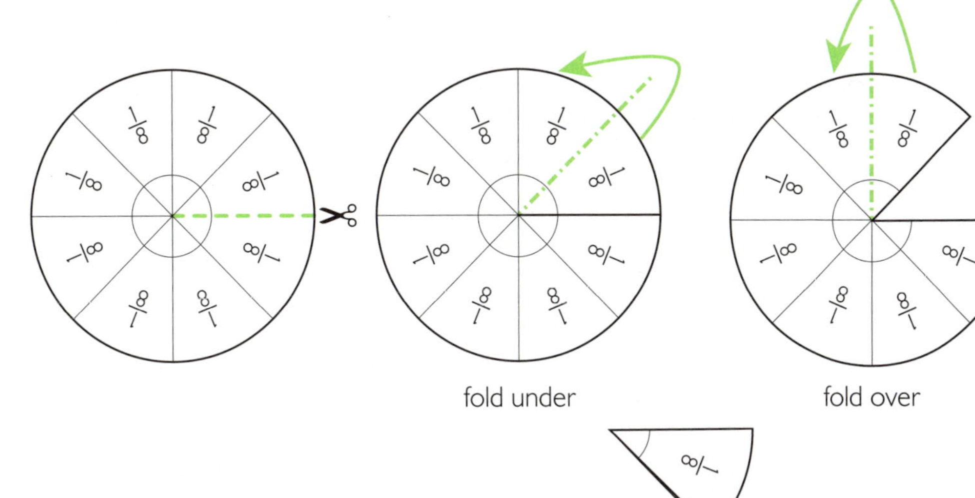
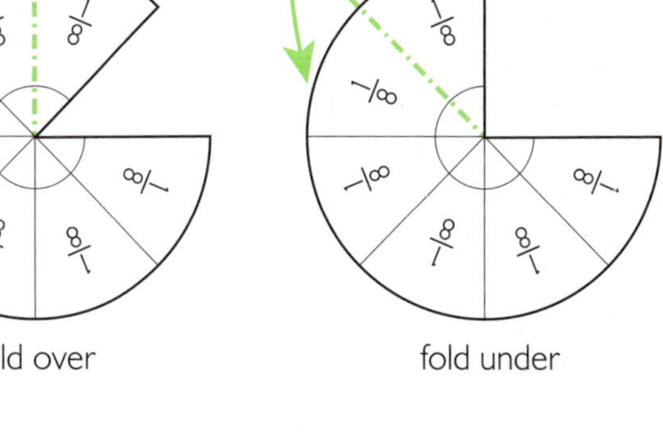

▲ *The students can make a more useful tester by cutting and folding the tester from Blackline Master 10.*

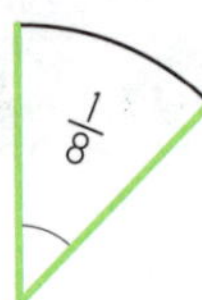

▲ *The students can use the tester to measure the shapes shown on Blackline Master 9.*

■ Materials
- Blackline Master 11 (page 69)
- Students' circular eighth-angle testers from the previous activity
- Students' copies of Blackline Master 9 from the previous activity

■ Materials
- Blackline Master 10 (page 68)
- Blackline Master 11 (page 69)
- Blackline Master 12 (page 70)
- Photocopy paper — 3 different colors (e.g. yellow, blue, green) — use a different color for each blackline master
- Blu-Tack
- Scissors — 1 pair for each pair of students

---

3. The students can then fold and unfold their testers to record different angles. Each student can then measure and mark the sizes of the interior angles of the shapes on Blackline Master 9. Review where the angle arms and vertex of the tester are and how they must be aligned with the angle arms and vertex in the corners of the shapes. Some of the angles cannot be measured using whole eighths. Tell the students to leave these blank. The worksheet will also be used for the next activity.

8. Multi-Angle Tester 2

Preparation
1. Make enough copies of Blackline Master 11 on color paper so that there is a tester for each student.

2. Make an enlarged copy of one tester from Blackline Master 11 for demonstration.

Activity
1. Instruct each student to cut out and fold up one of the twelfth-angle testers from Blackline Master 11. Say, *Compare a single eighth angle with a twelfth angle. What do you notice?* (The twelfth has a smaller amount of turn or opening between the angle arms.)

2. Ask, *What other amounts of a full turn can you measure with this tester beside twelfths?* Bring out the fact taht the students can also measure a half, a third, a quarter, and a sixth of a full turn by using combinations of twelfths.

3. Instruct the students to re-examine Blackline Master 9 from the previous activity. They can use their twelfth-angle testers to measure the interior angles that they were unable to measure exactly. Then have the students check the interior angles they measured with the eighth-angle tester to see if any of those can also be measured using the twelfth-angle tester.

9. Angle Wall

Preparation
1. Make two copies of Blackline Master 10 (eighth-angle testers) so that there is one tester each for four pairs of students.

2. Make three copies of Blackline Master 11 (twelfth-angle testers) so that there is one tester each for six pairs of students.

3. Make one copy of Blackline Master 12 (quarter-angle testers) so that there is one tester each for two pairs of students.

Activity

1. Instruct each pair of students to cut the testers they are given to produce the angles nominated. Arrange the students so that each pair works on one of the sets of angles listed below. For example, direct one pair of students to cut an eighth-angle tester to produce a one-eighth angle and a seven-eighths angle. Direct another pair to make a two-eighths angle and a six-eighths angle.

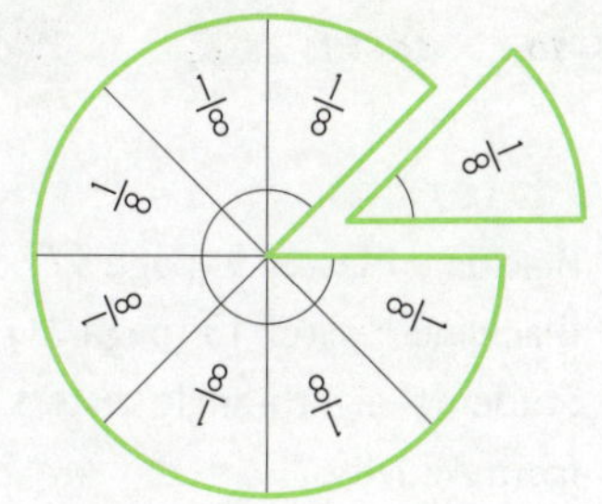

▲ *One pair of students will produce a seven-eighths angle and a one-eighth angle by cutting a tester.*

Quarters	
$\frac{1}{4}$ and $\frac{3}{4}$	$\frac{2}{4}$

Eighths	
$\frac{1}{8}$ and $\frac{7}{8}$	$\frac{2}{8}$ and $\frac{6}{8}$
$\frac{3}{8}$ and $\frac{5}{8}$	$\frac{4}{8}$

Twelfths	
$\frac{1}{12}$ and $\frac{11}{12}$	$\frac{2}{12}$ and $\frac{10}{12}$
$\frac{3}{12}$ and $\frac{9}{12}$	$\frac{4}{12}$ and $\frac{8}{12}$
$\frac{5}{12}$ and $\frac{7}{12}$	$\frac{6}{12}$

▲ *Divide the students into pairs to produce these sets of angles.*

2. After cutting the testers to show the required angles, give students who did not cut a tester the task of sticking them to the wall for display. Discuss the order with the students beforehand. Instruct students to orient the angle pieces so that it is easy to compare them. The entire display needs to show the angles from smallest to largest.

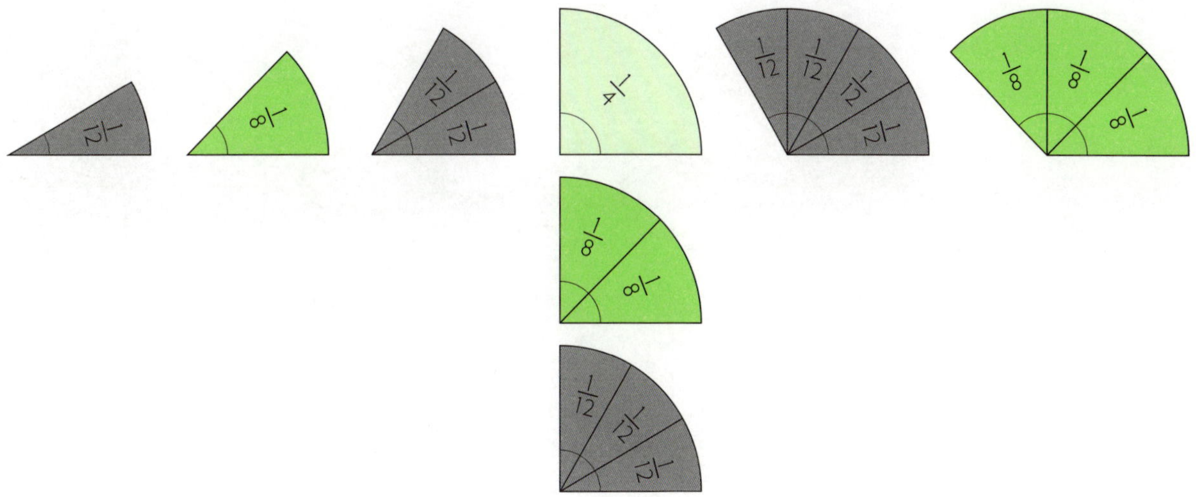

▲ *Have the students arrange the angles from smallest to largest. Here is one example of a display showing angles from a one-twelfth angle to a three-eighths angle.*

■ Materials

- Blackline Master 9 (page 67)
- Blackline Master 13 (page 71)
- Students' eighth-angle testers from Activity 7
- Students' twelfth-angle testers from Activity 8
- Blank transparency sheets — 1 for each student
- Overhead projector

10. Twenty-Fourth-Angle Tester

Preparation

1. Make a copy of Blackline Master 9 for each student.

2. Make an overhead transparency (OHT) of Blackline Master 9.

3. Copy Blackline Master 13 onto an OHT.

4. Make enough copies of Blackline Master 13 onto OHT sheets so that every student has a twenty-fourth-angle tester.

Activity

1. Lead a discussion to remind the student that they had to use two different testers to measure the angles on Blackline Master 9. Bring out the concept that it would be useful to have just *one* tester that could measure all of the angles.

2. Say, *We have to divide a tester into a larger number of equal angles. The number must be able to be divided into 8 equal parts and also 12 equal parts at the same time. What could that number be?* Suggest that the students list the multiples of 8 and 12 and see if there are numbers in each list that are the same. As the students start to list the multiples of 8 and 12 they will quickly discover that 24 is a common multiple.

3. Display the twenty-fourth-angle tester OHT on the overhead projector. Point out that there are twenty-four equal angles shown. Ask, *If we divide this tester into eight equal parts, how many angles are in each part?* (Three.) *Twelve equal parts?* (Two.) *What other equal parts can you divide it into?* (Two, three, four, and six.)

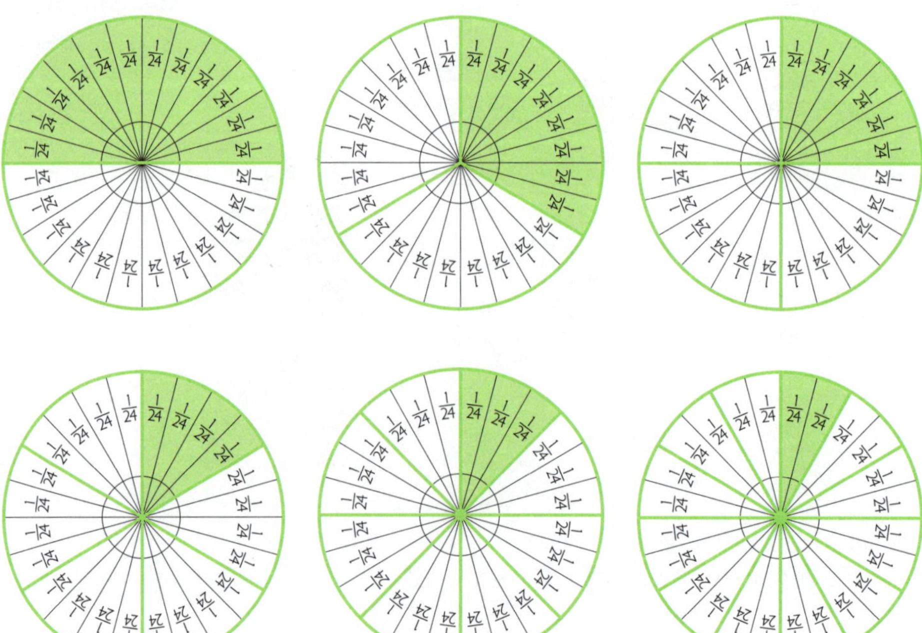

▲ *The students will discover that a twenty-fourth-angle tester can be divided into halves, thirds, quarters, sixths, eighths, and twelfths.*

Did You Know?

Analog clock faces can be used to show two different ways of dividing a full turn into equal amounts. The hour marks show twelfth angles, while the minute marks show sixtieth angles.

4. Distribute the twenty-fourth-angle tester to the students and discuss the advantages of having a flat, see-through tester (e.g. eliminating folding and unfolding). Have them compare this new tester with their other testers. Say, *Look at all of your different testers. What pattern do you notice about the opening between the angle arms?* (The more angles there are, the smaller the opening between the angle arms of each angle.) *How many twenty-fourth angles are equal to an eighth? Three-eighths? Five-twelfths?*

5. Have the students examine Blackline Master 9 and use their twenty-fourth-angle tester to measure and label the interior angles of the shapes.

11. Twenty-Fourth Turns

Preparation
No preparation is required.

Activity

1. It might be helpful at this stage to highlight that the number of equal corners around a point can also be expressed as "amounts of turn". Display the overhead transparency and place the geostrips in a closed position on top of the tester. Place the fastener on the central point of the tester, with the strips aligned with one of the angle arms on the tester. As you hold one geostrip and turn the other, ask students to describe the amount of *turn* that the strip is making (for example, ten twenty-fourths). Repeat this activity with different amounts of turn.

2. Encourage the students to use their testers to measure the maximum amount of turn that different objects in the room have. For example, the blades of a pair of scissors may turn nine twenty-fourths of a full turn.

■ Materials
- Overhead transparency of the twenty-fourth-angle tester from the previous activity
- 2 geostrips with fastener
- Overhead projector and blank transparency sheet

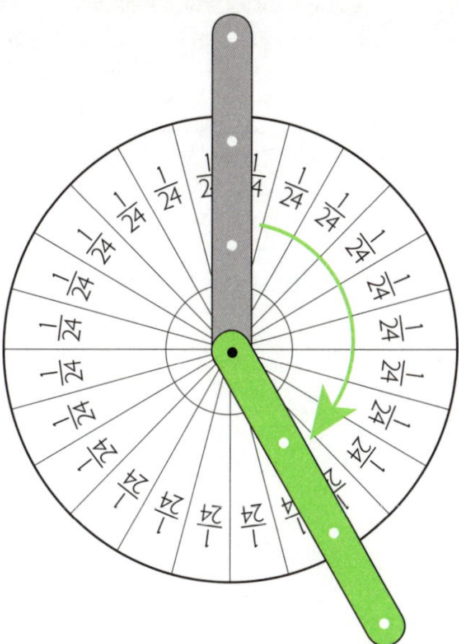

▲ *Highlight how the twenty-fourth-angle tester also measures amounts of turn.*

12. Examining Fractions and Factors

■ Materials

- Blackline Master 14 (page 72)
- Overhead transparency of the twenty-fourth-angle tester from the previous activity
- Overhead projector and blank transparency sheet
- Calculator — 1 for each student

Did You Know?

The Babylonians observed that the sun completed a cycle through the sky approximately every 360 days. Perhaps because of this fact, and because 360 has many factors, one full turn around a point was also divided into 360 equal parts.

Did You Know?

A number that can be multiplied a certain number of times to produce a given number is a factor of that number.

Some students may think that the larger the number, the more factors it will have. They may be interested to learn that 360 has even more factors than 1000.

Preparation

Make an overhead transparency (OHT) of Blackline Master 14.

Activity

1. Display the OHT and discuss the shapes shown. Bring out the fact that the shapes are regular polygons. Lay the twenty-fourth-angle tester OHT on top of one of the corners of the square and invite a volunteer to name the number of twenty-fourth angles in the interior angle of the corner. Repeat with the triangle.

2. Place the tester over the pentagon and bring out the fact that you cannot use the tester to measure the interior angles of the pentagon exactly.

3. Elicit the students' agreement that it would be useful to have a tester that could measure the interior angles of such a basic shape as a regular pentagon. Discuss how the tester will need to have smaller units of measure but should still be as useful as the existing testers. That is, the tester should still be able to measure a half, third, quarter, sixth, eighth, twelfth, and twenty-fourth of a full turn around a point.

4. Ask the students to suggest numbers that they think might be useful. Some students may immediately suggest 100, among other numbers. Write three suggestions on the board along with 360 and have the students find the factors of the numbers. The factors should be whole numbers.

5. Discuss how 360 is one of the standard number of divisions that is used to break a full turn into equal amounts. There are a few reasons for this, one being that there are so many ways of dividing 360 into equal parts.

Factors of 48	
1	48
2	24
3	16
4	12
6	8

Factors of 100	
1	100
2	50
4	25
5	20
10	10

Factors of 1000	
1	1000
2	500
4	250
5	200
8	125
10	100
20	50
25	40

Factors of 360	
1	360
2	180
3	120
4	90
5	72
6	60
8	45
9	40
10	36
12	30
15	24
18	20

▲ *Students can calculate the factors of a number by dividing the number by one, then two, and so on until the numbers start to repeat.*

13. Using a 360° Protractor

Preparation

1. Make a copy of Blackline Master 14 for each student.

2. Copy Blackline Master 15 onto blank transparency sheets so that every student has a 360° protractor.

3. Make an overhead transparency (OHT) of Blackline Master 15 for demonstration.

4. Make a copy of Blackline Master 16 for each student.

Activity

1. Display the OHT of the protractor. Discuss how it shows 360 equal angles. Say, *Each of these angles is one three-hundred-and-sixtieth of a full turn. 360 of these tiny angles fit around a point. Instead of calling one of these tiny fractions a "three-hundred-and-sixtieth angle", we use the word "degree".* Introduce the degree symbol (°) at this point.

2. Distribute the students' copies of the protractor and have them compare it with their old testers. They may wish to trim around the circle to make it neat. Introduce the term "protractor" as another word for an angle tester.

3. By this stage the students should be able to work comfortably with abstract representations of angles. Have them each draw a large capital "V". Do similarly on a blank overhead transparency and go through the instructions on Blackline Master 16 with the students.

4. Afterward, the students can measure the interior angles of the shapes on Blackline Master 14. They will discover that the angles of the triangle are each 60°; the square, 90°; the pentagon, 108°; and the hexagon, 120°.

■ Materials

- Blackline Master 14 (page 72)
- Blackline Master 15 (page 73)
- Blackline Master 16 (page 74)
- The eighth-, twelfth-, and twenty-fourth-angle testers from Activity 12
- Overhead transparency sheets — 2 for each student
- Overhead projector
- Overhead transparency pen
- Overhead transparency of Blackline Master 14 (from the previous activity)

Did You Know?

The word "protractor" comes from two Latin words — *pro*, meaning "forward", and *trahere*, meaning *"draw or drag"*. To use a protractor, it has to be imagined that a line is being dragged or turned from one position towards another.

Did You Know?

Regular polygons have all angles and all sides equal.

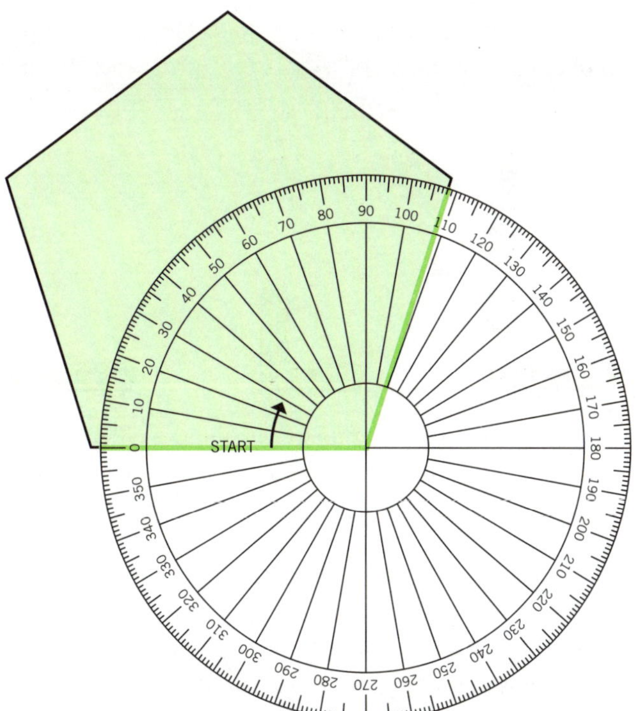

▲ *Students will find using the 360° protractors from Blackline Master 15 much easier than using standard protractors.*

 **angles with two arms**

These activities are designed for students to examine different types of situations that involve two angle arms. Challenge the students to find examples inside and outside of the classroom of objects or situations where two angle arms can be seen. If the students are comfortable measuring angles, they can also describe the magnitude of the angles they investigate in the activities below.

■ Materials

- Geostrips (3 long strips and 3 short strips) with fasteners

1. Long and Short Arms

Preparation

Attach the geostrips in pairs so that there are two long strips together, two short strips together, and one long and one short strip together.

Activity

1. Hold the long geostrips so that the two strips are aligned over one another. Hold one and rotate the other a quarter turn. Ask, *How much did I turn one of the strips?* (A quarter turn.) Choose an individual to hold the strips in this position for you.

2. Repeat the previous step with the pair of short geostrips. Say, *What do you notice about the amount of turn that both pairs of geostrips made?* (They both made a quarter turn.) *What do you notice about the size of the opening between the strips in each pair?* (They look the same.) *What do you notice about the length of the geostrips?* (One pair is longer than the other.)

3. Align the remaining pair of geostrips and turn one of them a quarter turn. Ask, *What amount of turn happened here?* (A quarter turn.) Note that the strips are different lengths and have a student hold the strips for display.

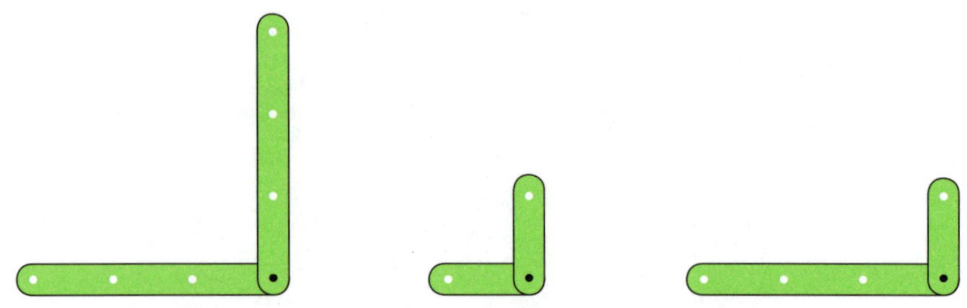

▲ *By comparing the angles that three different pairs of geostrips show, the students can describe how the length of the angle arms does not affect the amount of opening between the arms.*

4. Point to the three pairs of geostrips and ask, *Did one of the strips in each pair turn more than the others?* (No, they all turned the same amount.) *Which pair of strips has the largest opening between the arms?* (None. They all have the same opening.) *We can think of the geostrips as angle arms. Does the length of the angle arms affect the amount of turn that they can make?* (No.) *Do the angle arms have to be the same length?* (No.)

5. Have the students find three examples in the classroom where they can see pairs of angle arms that are the same length and three pairs that are different lengths. Then challenge the class to find three examples of pairs of angle arms that show the same amount of turn or opening but are different lengths to the other pair.

2. Walking Angles

Preparation

Use chalk or masking tape to draw the following figures on a large open space. The combined length of the lines in each figure should be enough that a number of students can walk along the lines at the same time.

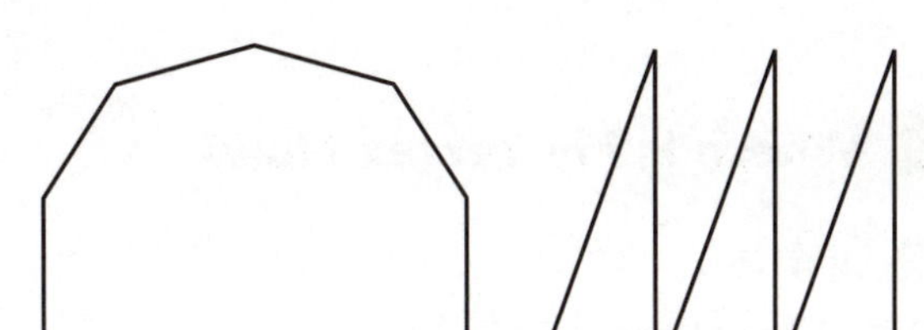

■ **Materials**
• Chalk or masking tape

Activity

1. Arrange the students into three equal groups and have each group walk along one of the figures you drew. Rotate the groups so that each group walks along each figure.

2. Afterward, discuss the types of turns that the students made as they walked along the lines. Allow the students to use their own words to describe the types of turns they made. Bring out the fact that on one figure they made quarter turns, on the second they turned a lot less, while on the third they almost made full turns.

3. Invite volunteers to identify pairs of angle arms and vertices in each of the figures.

geo Other investigations on following paths can be found in *Making Moves*.

3. Rounded Corners

Preparation

No preparation is required.

Activity

1. On the board, draw the figure shown on the right and have one student from each group draw a similar diagram. Highlight that the figure can be seen as a straight line that has been bent, like a garden hose.

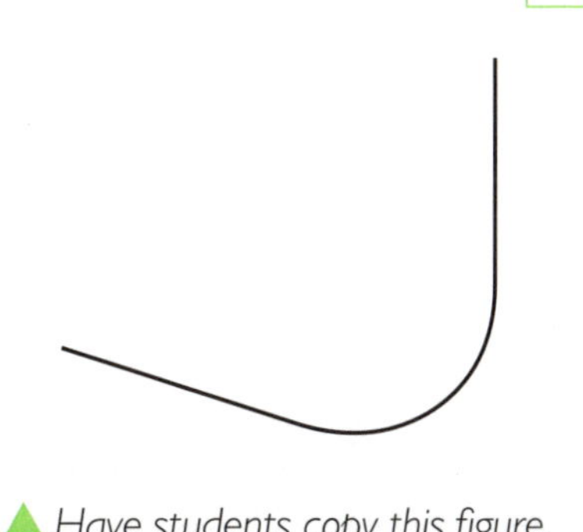

▲ *Have students copy this figure.*

■ **Materials**
None

2. Say, *This figure shows a type of corner. But where are the angle arms and vertex?* Bring out the fact the straight lines are the arms and the vertex is the entire length of the bend. Point out that such a vertex is not is very precise, as there is no clear place where the arms join together.

3. Say, *Ignore the bend and imagine where the straight lines would meet if they continued. Draw what this would look like.* Discuss the students' results and change the board diagram so it looks like that shown on the right.

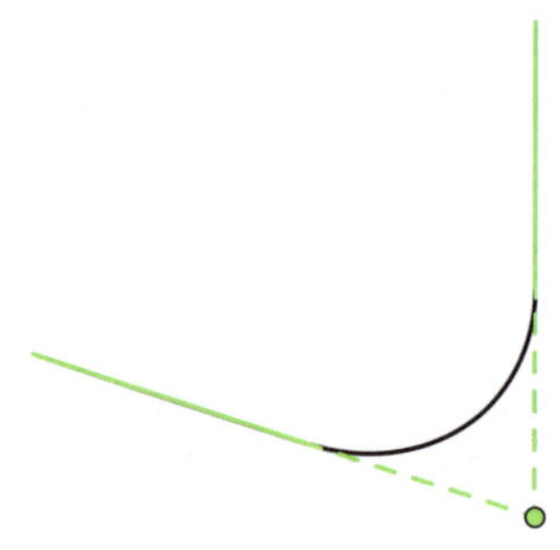

▲ *Demonstrate how to find the vertex of angle arms in a rounded corner by extending the arms until they meet.*

■ Materials

None

Did You Know?

A "vertex" is the point where two or more line segments join or intersect on a 2D shape, or where three or more edges meet on a 3D shape. Also, the point where two or more angle arms intersect.

4. Where Is My Vertex Now?

Preparation

No preparation is required.

Activity

1. Instruct each student to place their hands together and stretch their arms straight above their heads. Direct each student to move both arms to show an opening between their arms.

2. Ask, *If our arms are like angle arms, where is the vertex?* Encourage the students to imagine their arms continuing in a straight line through their bodies. Explain that the vertex is where they would eventually meet.

▲ *Challenge the students to picture where the vertex will be if they extend the lines made by their arms.*

3. Do a few examples as a class then each student can work with a partner. One student from each pair can arrange his or her arms and the other student can identify and record the position of the vertex.

5. Surfaces

10-12

Preparation
No preparation is required.

Activity

1. Have the students fold a sheet of paper in half, then challenge them to identify the angle arms and vertices that they can see. One pair of arms are those that form the edges of the paper. The angle vertex then is where the two edges meet at the fold.

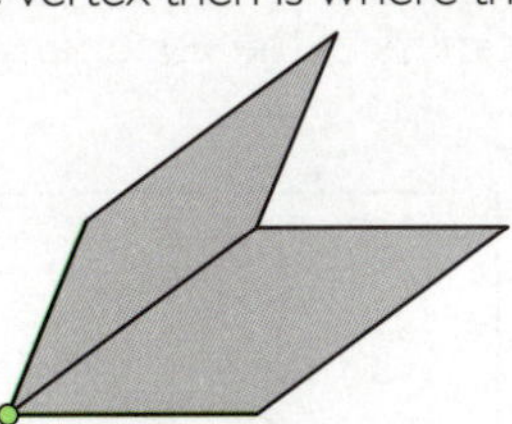

▲ *Very thin angle arms can be seen on the edge of a folded sheet of paper.*

2. Say, *We can also view the angle arms as being spread out over the surface of the paper, like smearing a straight line of butter across some bread. The two halves of the paper can then be seen as two very wide angle arms, with the entire length of the crease between them as the vertex.* Encourage the students to identify similar examples of this concept such as the faces of 3D shapes, steps, and between the pages of a book.

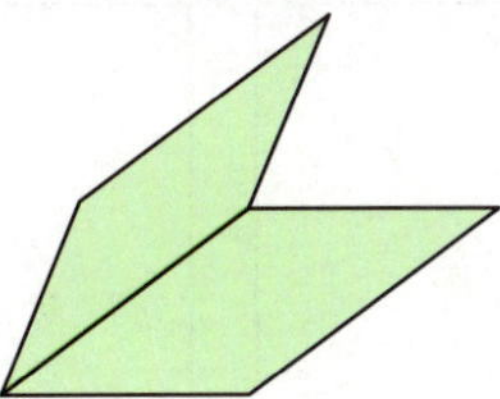

▲ *The two parts of a folded sheet of paper can be seen as very thin but wide angle arms, with the crease being the vertex.*

■ Materials
None

Did You Know?
A "dihedral angle" is the angle between two planes, like two faces on a cube.

A "solid angle" is found where three or more planes meet, like the vertex of a cube. The size of this angle is measured in steradians instead of degrees.

A "vertex" is the point where two or more line segments join or intersect on a 2D shape, where three or more edges meet on a 3D shape or, the point where two or more angle arms intersect.

6. Thick Arms

10-12

Preparation
No preparation is required.

Activity

1. Discuss the amount of turn that a forearm and an upper arm can make together.

2. Have each student work with a partner to trace the outline of their bent upper arm and forearm onto a sheet of paper.

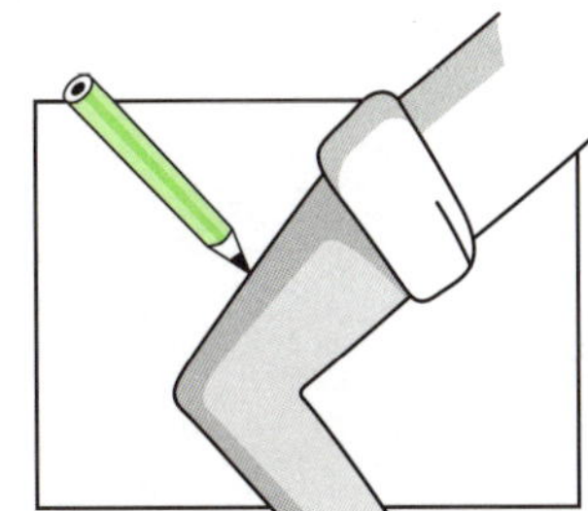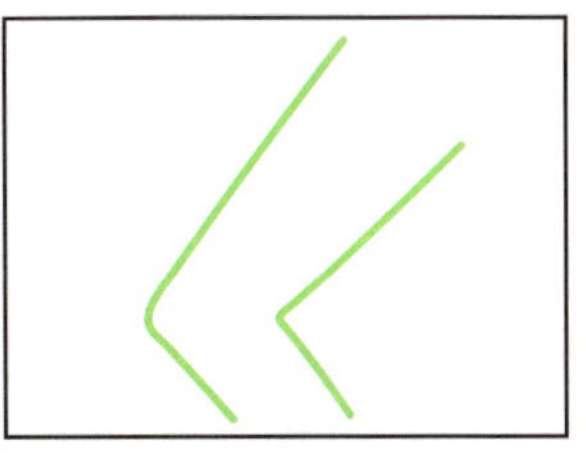

▲ *Students can investigate where the vertex is on their arms.*

■ Materials
None

3. Elicit the students' agreement that they can use their arms to make openings of different amounts, and that their upper arm and forearm are like angle arms. Challenge the students to discuss their ideas with their partner then identify and record the vertex on their drawings.

4. Discuss the students' drawings. Bring out the fact that the thicker the angle arms are, the harder it becomes to work out where the vertex actually is. When looking at the vertex made by a forearm and an upper arm, there are at least four different approaches to locating a vertex. One approach is to see the vertex as being as wide as the thickness of the upper arm and forearm (see below).

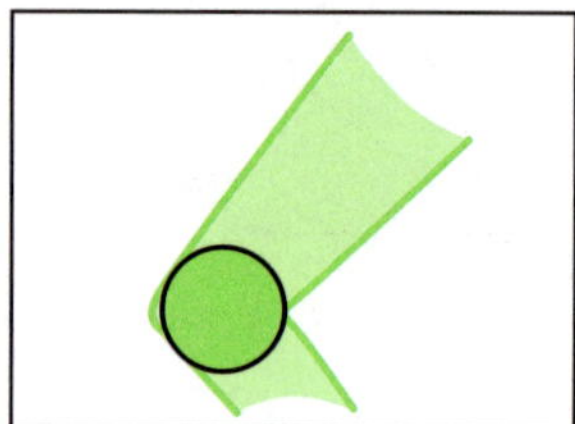

Another approach is to consider the outlines made by the inner arm or the outer arm and locate the vertex that way (see below).

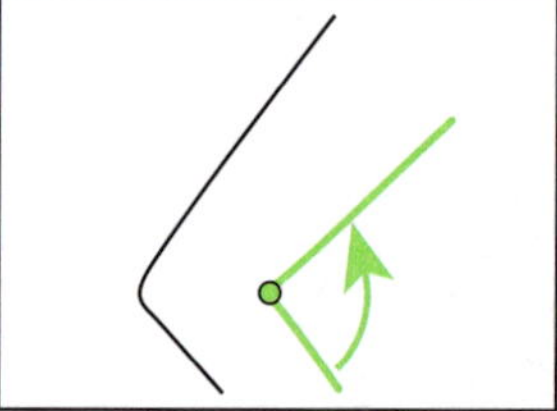 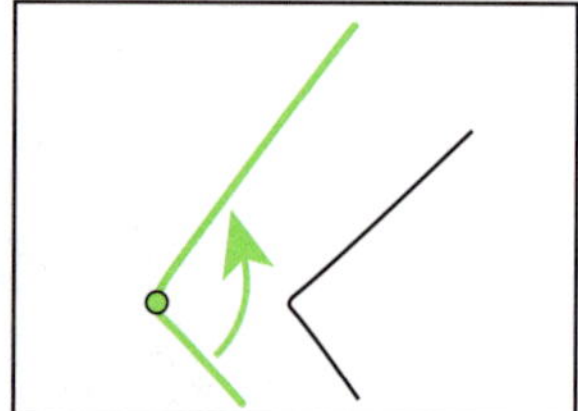

The most accurate way is to consider lines running between the inner arm outline and the outer arm outline. In this case, the vertex is where these two lines meet (see below).

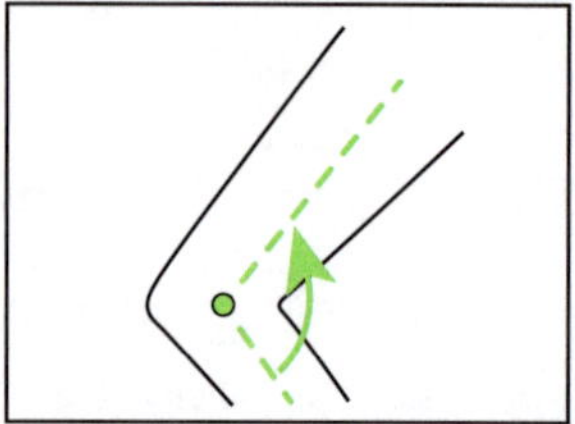

5. Encourage the students to investigate other examples of "thick arms" in the classroom.

angles with one or zero arms

Identifying and measuring angles is harder when only one or none of the angle arms can be seen. The activities below are designed to help the students develop the ability to picture the missing arms. Ask the students to find examples of these types of angle contexts as they are examined.

1. Signposts and Flagpoles

Preparation

Distribute the playdough and ask students to flatten it into a flat surface about 2 cm thick.

Activity

1. Say, *Stick a pencil into the playdough to represent a flagpole. Where are the vertex and two angle arms when you look at a flagpole in the ground?*

2. Using the materials, the students will be able to identify the "flagpole" as being one arm but may not be able to describe the other. This is because when the pole is upright it can be any imaginary line on the ground that ends at the base of the pole (which is the vertex). If the pole is tilted then the imaginary line must be aligned directly underneath it.

3. Instruct the students to identify angle arms when the pencil is upright and also when it is leaning. Afterward, have the students to identify the angle arms in similar examples inside and outside the classroom.

■ Materials

- Playdough or similar (tennis ball-sized amount) — 1 for each group of students

Did You Know?

A "vertex" is the point where two or more line segments join or intersect on a 2D shape, where three or more edges meet on a 3D shape, or the point where two or more angle arms intersect.

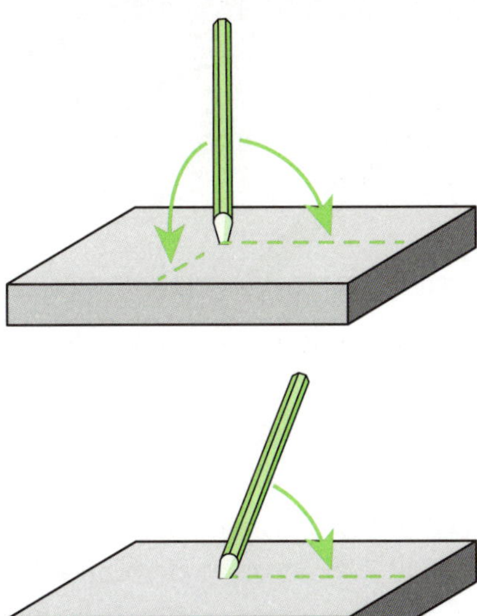

▲ *Ask the students to consider where the angle arms would be in an upright flagpole and a leaning one.*

■ Materials

- 2 geostrips and fastener
- Blu-Tack

2. Gauges

Preparation

1. Fasten the geostrips at one end.

2. On the board, draw the fuel gauge shown below. The diameter of the semicircle should be two ruler lengths.

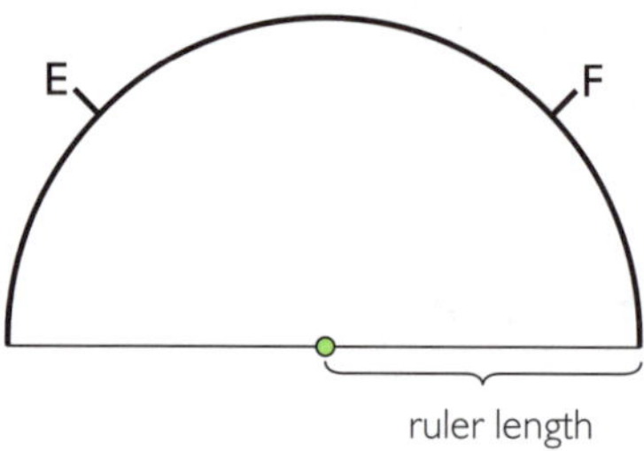

Activity

1. Align the geostrips over one another. Hold one strip and turn the other a quarter turn. Have the students draw a picture to represent the amount of turn or opening that the strip made.

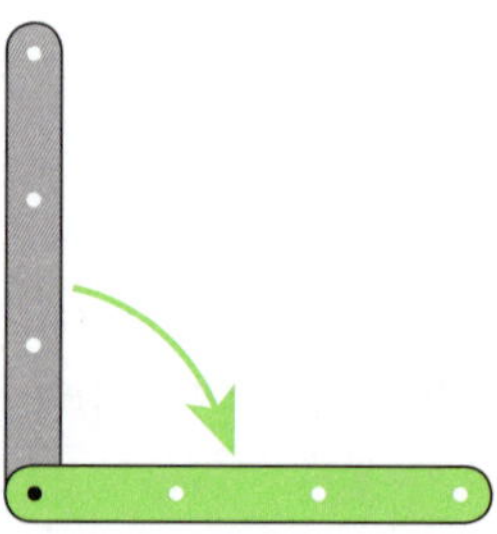

2. Discuss the fuel gauges in cars including what the "E" (empty) and "F" (full) stand for. Place a ruler on the gauge so that it points to "E" (use the Blu-Tack to keep it in place). Have the students sketch the fuel gauge.

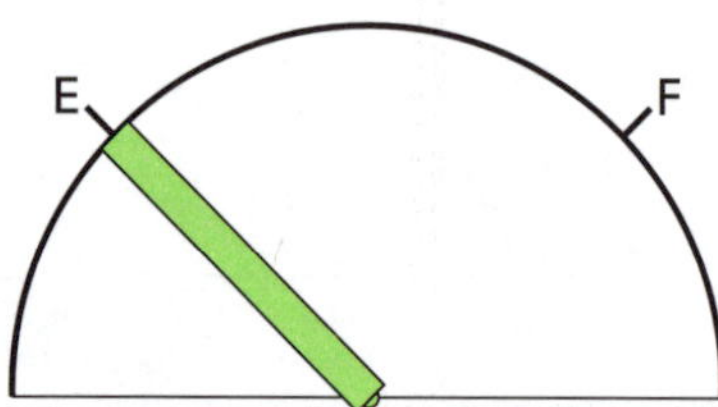

▲ *Have the students sketch the model of a fuel gauge.*

3. Say, *The needle (i.e. the ruler) in this fuel gauge is pointing to "empty". When we put fuel in the car, the needle moves toward "full".* Move the ruler so that it rests on "F". Ask, *How much of a turn did the needle make?* Discuss the students' answers, encouraging language such as "more than" and "less than". Ask, *How did you figure out how far the needle turned?* One suggestion may be to imagine where the starting position was, compared to the finishing position. Define the vertex the needle rotated around.

Did You Know?

Other examples of where there is only one angle arm include swings and boom gates.

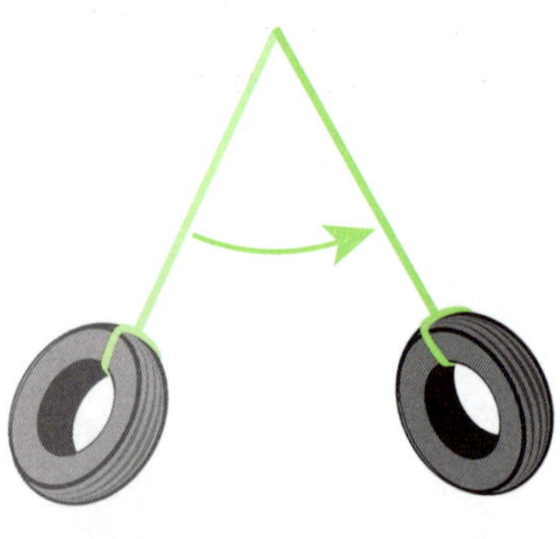

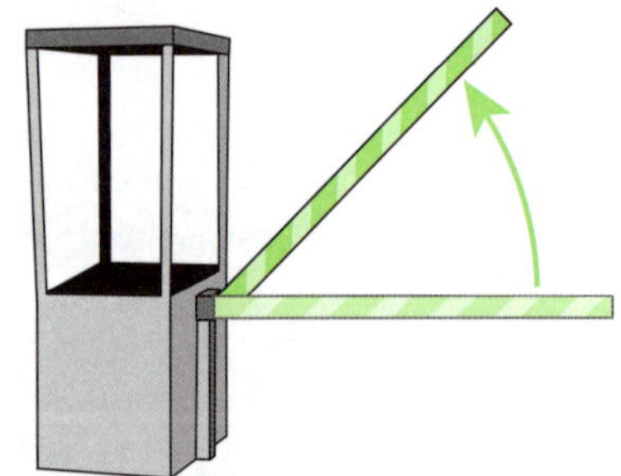

4. Say, *With the geostrips, the angle arms started in the same position. When we moved one of the strips we saw the starting and finishing positions at the same time. How is this different from the fuel gauge?* (The starting or finishing position can be seen, but not both at the same time.) Say, *Try to draw a picture that shows the amount of turn that the needle moved.*

5. After the students draw their picture, discuss their ideas. Show them how to draw the starting position of the needle using dotted lines, and then draw the finishing position with solid lines. An arrow can be used to indicate the direction the needle turned.

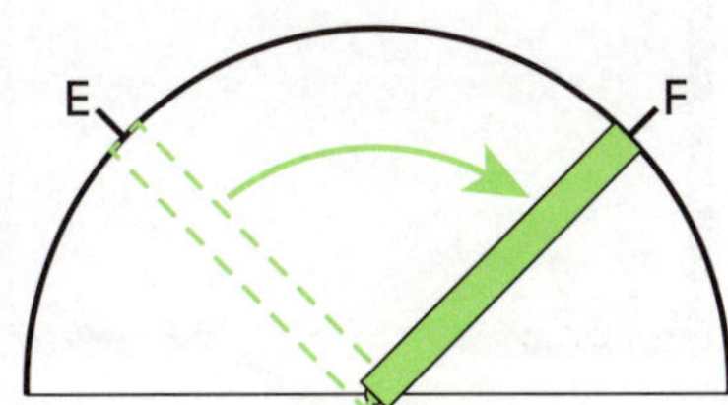

▲ *To help the students see the amount of turn that an angle arm makes, ask them to sketch the starting position and finishing position.*

3. Doors

Preparation

Cut a door into the demonstration box to create a door (see below).

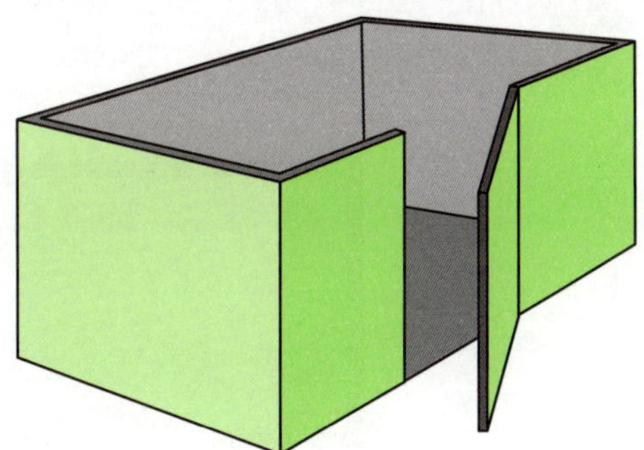

■ Materials
- Cardboard box without a lid — 1 for each group of students and 1 for demonstration
- Scissors — 1 pair for each group of students
- Pattern blocks

Activity

1. Discuss how the classroom door opens and closes by turning on its hinges. Show the demonstration box and explain that it is like a room with a door. Each group can cut a similar "door" in their own box.

2. Say, *Consider what you know about changeable angle arms. What aspect of your model is a vertex?* In addition to the corners of the room, ensure the students identify the hinged section of the door as being a vertex.

3. Open the door of the real-life classroom about a one-eighth (45°) turn from the closed position. Instruct each group to show the same amount of opening on their model's door.

4. Say, *The door is an angle arm and where the hinges are is a vertex. Where is the second angle arm?* Some students may suggest that the wall connected to the door via the hinge is the second angle arm. The second arm is actually the closed position of the door. Fortunately, the closed starting position of many real doors is marked at the bottom of the door frame (the sill) by a wood, metal, or plastic strip. On the model, it is the edge of the box from where the bottom of the door was cut. To help the students see the two angle arms clearly, ask them to take turns to look directly down upon the model room and door.

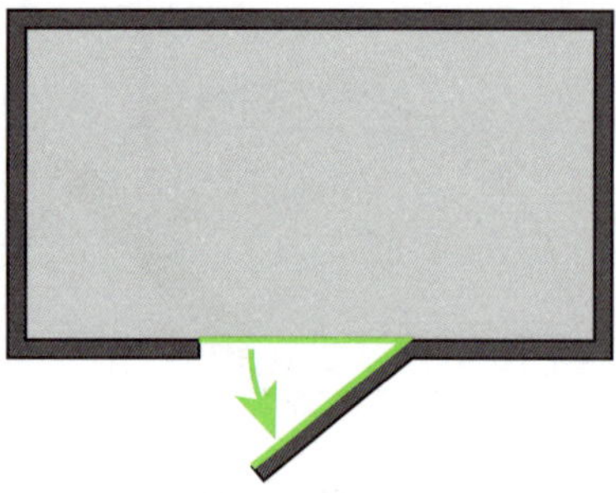

▲ *Viewing a model of a doorway from above will help the students identify the angle arms.*

5. Have the students identify one corner of a triangular pattern block. Challenge the groups to estimate the opening of the model door to match the size of the pattern block corner, without physically inserting the block. Once they have done that, have them record the outline made by the door and the sill on paper, as shown below. The pattern block corner can then be directly compared with the resulting angle arms. Repeat with angles from other pattern blocks.

▲ *Challenge the students to estimate the door opening and match the opening to a pattern block corner and record their estimation on a sheet of paper.*

4. Wheels

Preparation

1. Draw a circle on the smaller sheet of tagboard and cut out the resulting shape.

2. Attach the shape through the center to the middle of the larger sheet of tagboard with the fastener.

Activity

1. Show the students the spinner. Spin the shape an amount of turn no larger than a full turn. Ask, *How much of a full turn was made?* Discuss the difficulty of establishing the amount of turn.

2. Say, *Consider instances where you have figured out the amount of turn something made. What helped you measure the amount of turn?* Have the each student discuss his or her ideas with a partner.

3. Have each student make a spinner. Direct them to cut a sheet of paper in half. Then draw a circle on one half, cut the shape out and attach it to the remaining piece of paper with a fastener.

4. Introduce the method for recording amounts of turn. Draw a radius on the shape and mark the "12 o'clock" position on the backing board. Say, *Consider this radius as one angle arm and the center of the shape as the vertex. Then imagine the other angle arm running from the vertex to the mark on the backing board.* After rotation, the distance between the two angle arms can be examined.

■ **Materials**

- Sheet of tagboard or light card, approx. 20 cm square
- Sheet of tagboard or light card, approx. 25 cm square
- Drawing compass — 1 for each student and 1 for demonstration
- Scissors — 1 pair for each student and 1 for demonstration
- Fasteners — 1 for each student and 1 for demonstration

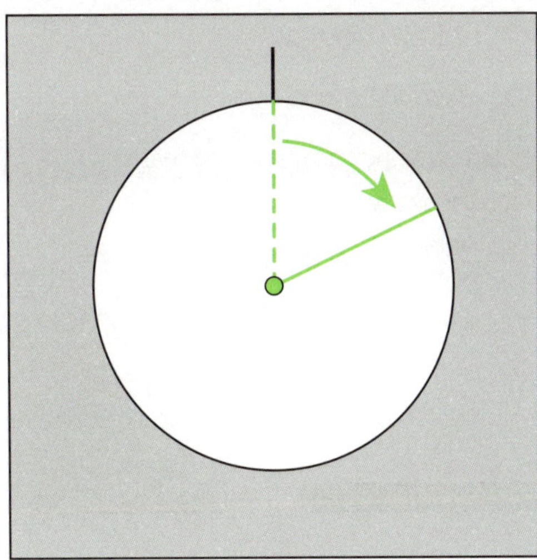

▲ *Seeing the amount of turn that is made by an object with no visible angle arms is easier if students mark the starting and finishing positions of a point on the object.*

Did You Know?

A radius is the line segment that joins the center of a circle with any point on its circumference. The plural of radius is "radii".

Did You Know?

Other objects that rotate like the disc include doorknobs, compact discs, ceiling fans, windmills, dials, and a range of wheels.

comparing angles

The following activites involve students comparing the sizes and contexts of different angles. As the students examine the ideas presented below, relate the ideas to what they already know about identifying and measuring angles. Other ideas for comparing angles can be found in Activity 3, "Comparative Angles", on page 48.

■ Materials

- Blackline Master 17 (page 75)

1. Fixed and Changeable

Preparation

1. Collect a variety of everyday objects and pictures of angles.

2. Make a copy of Blackline Master 17 for each student.

Activity

Challenge the students to find situations that display "fixed" and "changeable" angles (Blackline Master 17 provides some examples). Have them identify the angle arms and vertex in each instance. Challenge pairs of students to compare two different angles and explain the similarities and differences between them. Some examples for younger students to compare include:

- a pattern block corner with the corner of a desk
- a book corner with a pair of geostrips

Older students may be asked to compare:

- an opening door with an upper arm and a forearm
- a fuel gauge with a bent straw
- a doorknob with a wheel

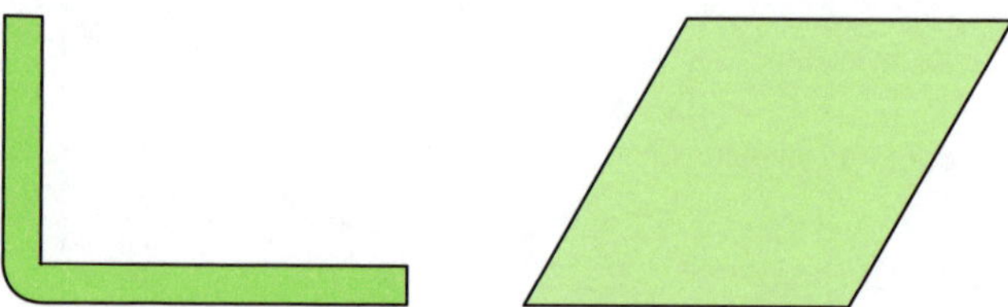

"The pattern block has four pairs of angle arms but the straw only has one pair."

▲ *Have the students compare two different contexts where angles can be observed.*

2. Absolute Angles

Preparation

Make a copy of Blackline Master 18 for each student.

■ **Materials**
• **Blackline Master 18** (page 76)

Activity

1. Some types of angles have particular names which students can find confusing because the names are somewhat abstract. As you discuss these angles with the students, provide many different examples and relate the new angle names to familiar ones. Distribute the copies of Blackline Master 18 to help students understand the origin of the names.

2. Introduce the term "right angle" to describe a quarter angle. Explain that the name has nothing to do with how the angle arms are oriented (there are no "left" angles) nor with correctness (there are no "wrong" angles).

▲ *Right angles.*

3. Introduce the term "straight angle" to describe a half angle. Students may find it easier to identify straight angles when seeing them as amounts of turn rather than as a type of "corner". The students may be interested to learn that every straight line is also a straight angle and that the vertex of the angle can be anywhere along the line.

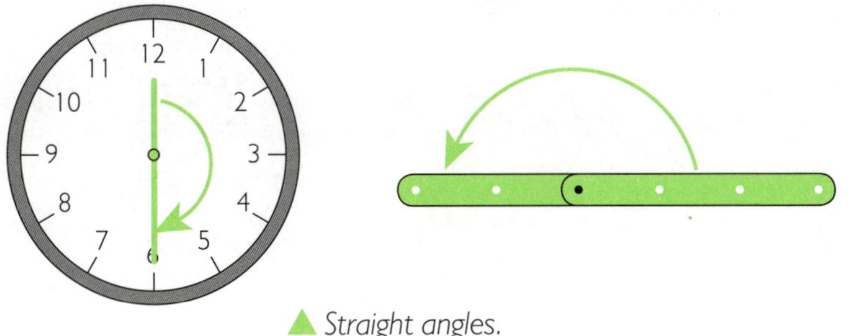

▲ *Straight angles.*

Did You Know?

A quarter angle is one-quarter of a complete turn around a point. It is equal to 90°.

A half angle is one-half of a complete turn around a point. It is equal to 180°.

A full angle is one complete turn around a point. It is equal to 360°.

4. There are a few names for a full angle. One is "perigon", another is "full turn", and a third is "revolution". These types of angles can only be described as an amount of turn around a point.

▲ *Revolutions or full turns.*

Materials

None

Did You Know?

Prior to the 20th century, two terms were used when describing types of interior angles of polygons. An angle less than a half angle was "salient", in the sense that these angles were prominent, or "stood out" from the shape (see below).

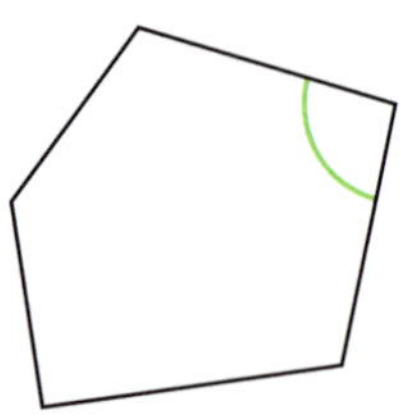

A reflex angle was "re-entrant", because the angle "re-entered" the shape (see below).

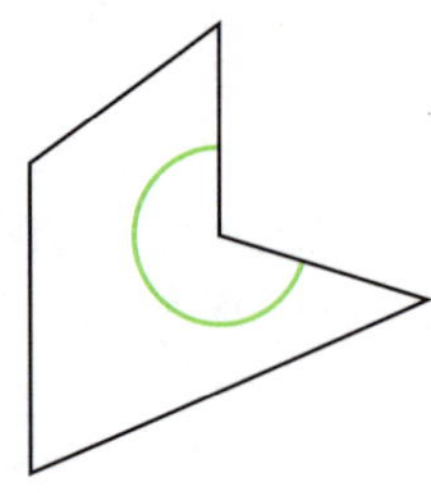

geo For more activities on investigating the properties of shapes, see *Paper Polygons* and *Plane Puzzles*.

3. Comparative Angles

Preparation

No preparation is required.

Activity

1. Say, *The names for the following types of angles are comparative. Unlike names such as "right" and "straight", these angles do not refer to an absolute amount of turn but are used as abbreviations for "less than", "more than", and "between".*

2. Introduce the term "acute" and explain that an acute angle is any angle less than a quarter angle (right angle).

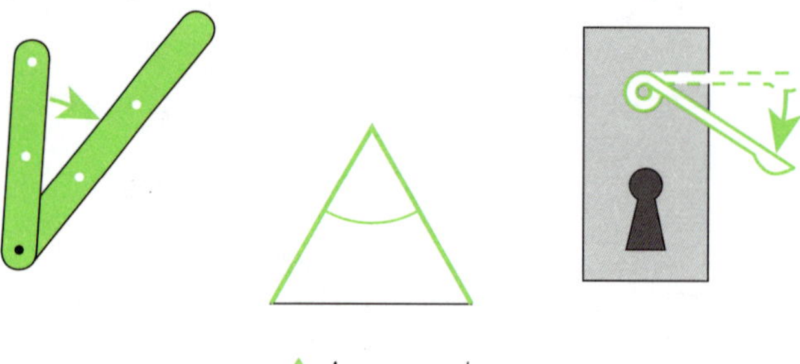

▲ *Acute angles.*

3. Introduce the term "obtuse" and explain that an obtuse angle is more than a quarter angle but less than a half angle (straight angle).

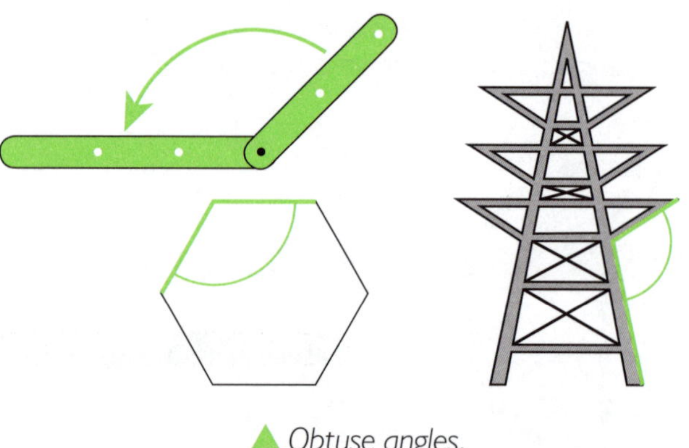

▲ *Obtuse angles.*

4. Last, introduce the term "reflex" to describe angles that are more than a half angle but less than a full turn.

▲ *Reflex angles.*

5. Have the students identify three examples of each type of angle.

4. Angles in Triangles

Preparation

Make a copy of Blackline Master 19 for each student.

Activity

1. For this activity, the students will need to make a quarter-angle tester (see below).

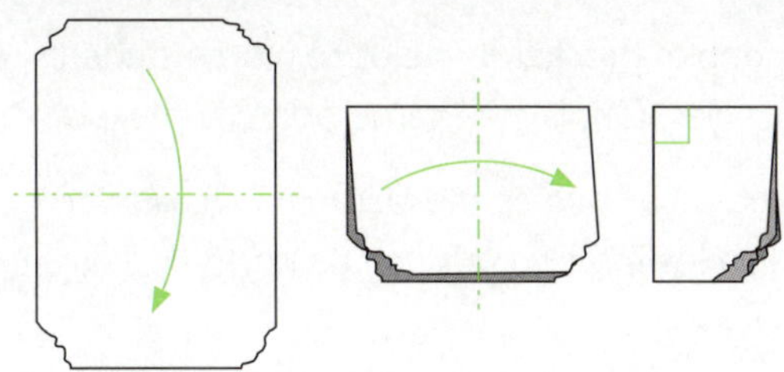

2. Have the students use their tester to examine the interior angles of the triangles on Blackline Master 19. Tell them to record the size of each angle using "A" for acute, "O" for obtuse, and "R" for right.

3. Discuss the students' answers and say, *Look at the angles in each of the triangles. What do all the triangles have in common?* (They all have at least two acute angles.)

4. Say, *We can name different types of triangles according to the size of their largest interior angle. An acute triangle has an acute angle as its largest angle. An obtuse triangle has an obtuse angle as its largest angle. A right triangle has a right angle as its largest angle. Use these names to label each triangle.*

■ Materials

- Blackline Master 19 (page 77)

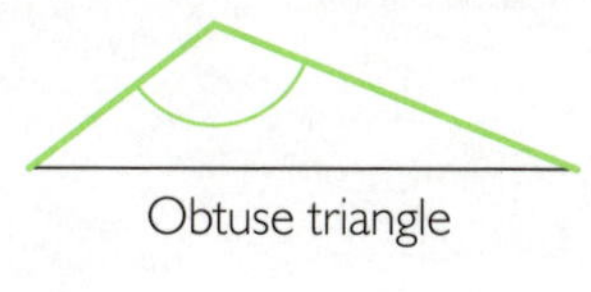

Obtuse triangle

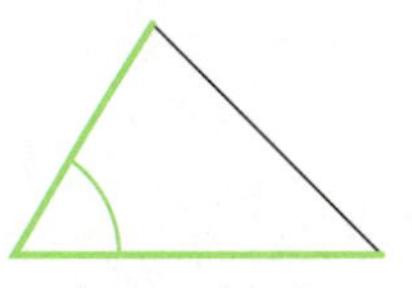

Acute triangle

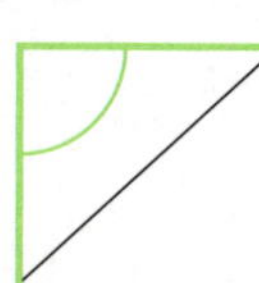

Right triangle

▲ *Explain that triangles can be named according to their largest interior angle.*

---

5. Angles in Quadrilaterals

Preparation

No preparation is required.

Activity

1. Direct the students to examine the quadrilaterals on the bottom of Blackline Master 19. Instruct them to label all the interior angles as in the previous activity. This time also using "RF" to record any reflex angles.

2. Ask, *What patterns can you see when you look at the type of angles in the shapes?* Have students work in pairs to discuss their thoughts.

3. Discuss the students' ideas and ask the following questions. During the class discussion, encourage the students to sketch diagrams to find exceptions or examples for a statement or question:

- *If a quadrilateral has three right angles, what will the fourth angle be?* (A right angle.)
- *If a shape has only two right angles, what will the other two be?* (One will be acute and one will be obtuse.)
- *Can you make a quadrilateral with two reflex angles? Why not?*
- *Can you name different types of quadrilaterals by the size of their largest angle like you can with triangles? What problems will there be?*

■ Materials

- Students' copies of Blackline Master 19 from the previous activity
- Students' quarter-angle testers from the previous activity

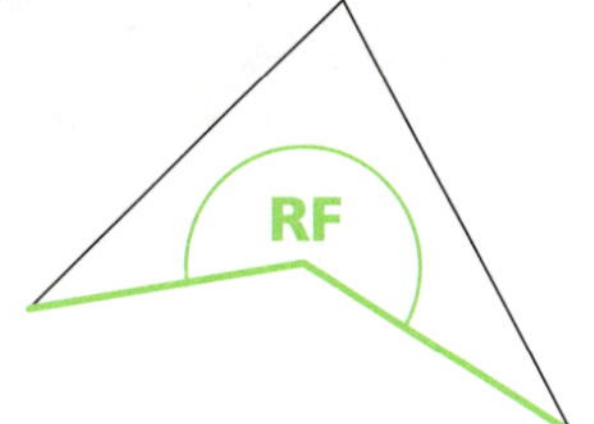

▲ *The students can mark reflex angles with the letters "RF".*

Did You Know?

An interior angle is an angle inside the perimeter of a polygon. Its angle arms are two adjacent sides of the polygon.

■ Materials

- Blackline Master 20 (page 78)
- Overhead projector and blank transparency sheet
- Standard die — 1 for each pair of students

6. Angle Chase

Preparation

1. Make a copy of the game board (Blackline Master 20) for each pair of students.

2. Copy Blackline Master 20 onto an overhead transparency.

Activity

1. A game for two people, this activity helps reinforce students' recognition of different types of angles. Display the game board and explain the rules as follows:

- *The object of the game is to be the first to reach all three targets. Both circular targets in the corners must be reached (in any order) before the players aim for the star in the middle.*

▲ *The targets.*

- *Each player starts on one of the lines already marked on the game board.*
- *Players take turns to draw a new line from the "free" end of the last line they drew. The new line can only be drawn to join one of the immediately surrounding dots.*
- *The type of angle the lines should form is determined by the roll of a die. (Refer to the key on the game board.) If a player cannot draw the angle, he or she misses a turn. A roll of six lets the player draw any type of angle from anywhere along his or her trail.*

 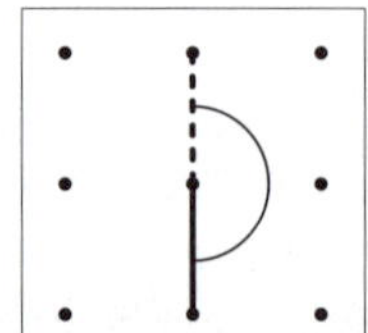 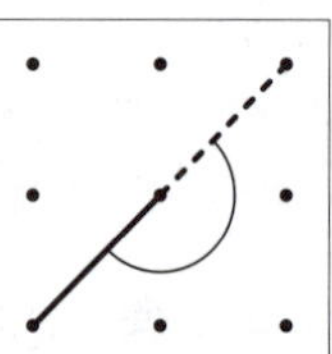

▲ *The angles that the students can draw are determined by the roll of a die.*

- *Each new line forms an angle with the line drawn previously by the same player and each player should mark the turn of the angle with an arc.*
- *Players cannot cross over their own trail.*

2. After the students have played the game, ask, *Which type of angle was most helpful to have? Least helpful? Why?* The replies will vary, but a sample response may include, *The reflex angle was helpful because there were many ways of showing it. The acute angle was less helpful because I started to double up on my trail.*

3. Discuss the representations of the types of angles shown on the game board. Point out that many other angles besides those shown can be acute, obtuse, or reflex. To avoid the students getting a stereotypical picture of what an acute angle is, for instance, have them each draw other examples.

4. It may be helpful to keep a laminated copy of the game board for the students to use at different times during the week. As an extension, have the students make up variations on the rules, such as only being able to turn in one particular direction (e.g. clockwise).

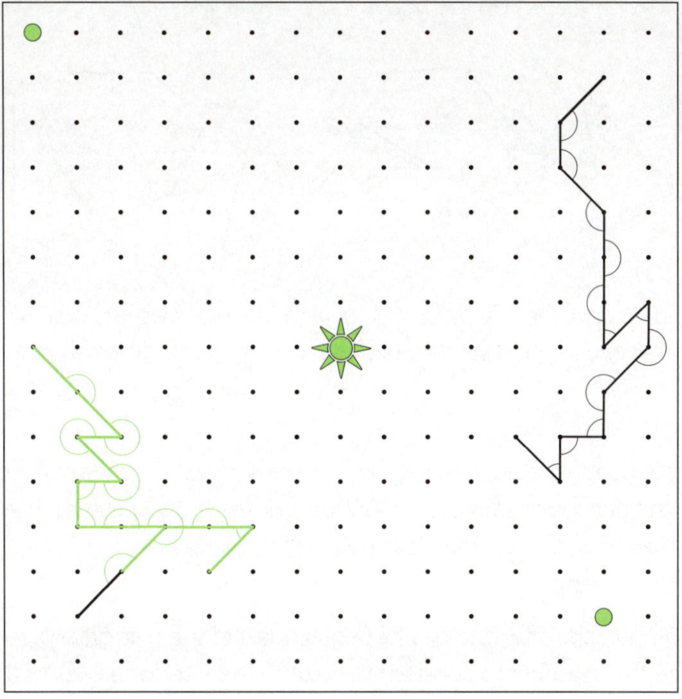

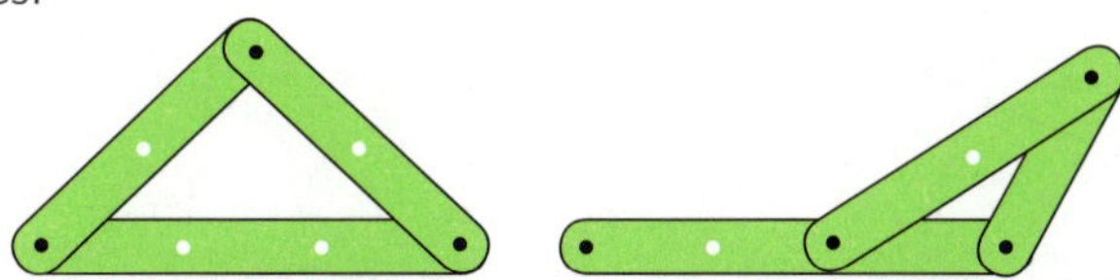
▲ *Playing "Angle Chase" will allow the students plenty of opportunity to use acute, right, straight, obtuse, and reflex angles.*

7. Geostrip Triangles

Preparation
No preparation is required.

Activity

1. Have the students select three geostrips and join them together to make a triangle. Direct the students to draw around the inside of the triangle.

2. Instruct the students to change the length of one of the geostrips and compare the new triangle with the outline of the previous one. Ask, *What happened to the size of the angles in the triangle?* (Some got bigger, some got smaller.) The students can repeat the sequence a number of times, noting the changes that take place.

3. Have the students swap the feature that is changed in the triangles. They can then change the size of an internal angle and observe the changes to the length of the triangles' sides.

▲ *The students will see that the size of the angles in a triangle change as the length of a side varies.*

Materials
*   **Geostrips and fasteners**

Materials

- Geostrips and fasteners or similar (e.g. drinking straws and pipe cleaners)

Did You Know?

Any polygon with four sides can be called a quadrilateral.

Did You Know?

"Re-entrant" is another word used to describe angles that are more than 180°.

8. Concave and Convex

Preparation

Draw the following shapes on the board.

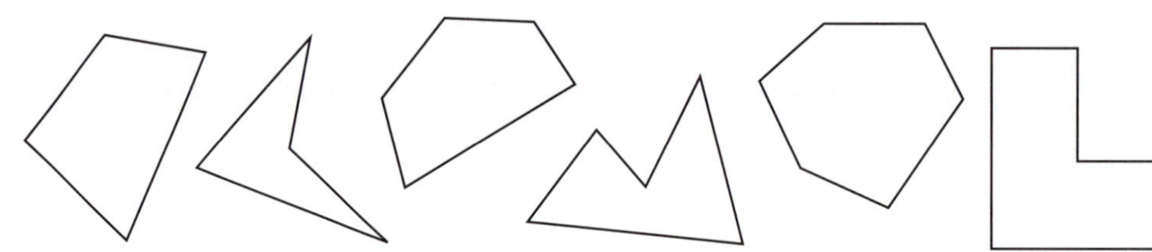

▲ *The students can identify the largest angle in each of these shapes to help them understand the difference between concave and convex shapes.*

Activity

1. Point to the quadrilaterals and ask, *What is the same about these two shapes?* (They both have four straight sides.) *What is different about them?* The students' answers may vary and include ideas such as the concave shape being "pointy" and "almost a triangle".

2. Repeat the questions for the pair of pentagons and the pair of hexagons. Encourage the students to consider the size of the interior angles of the shapes.

3. Ask, *What is the largest interior angle that you can see in each of the shapes? How can you describe the size of the angles? Are they acute, obtuse, or something else?*

4. Explain that some polygons can be called "concave" because they have at least one interior angle that is a reflex angle. Another way of describing this is to say that a concave shape looks like it has been "pushed in". "Convex" shapes, on the other hand, do not have any interior angles that are reflex angles.

5. Direct the students to use the geostrips to construct examples of concave and convex shapes.

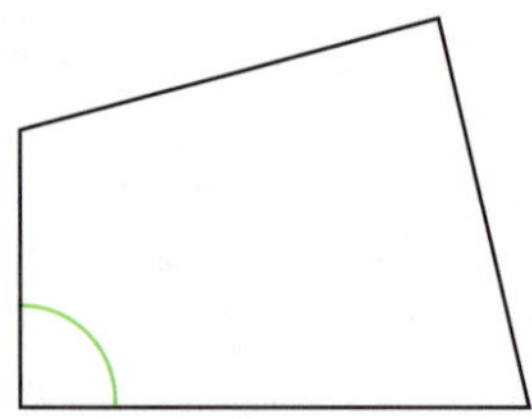

▲ *An interior angle lies inside a shape.*

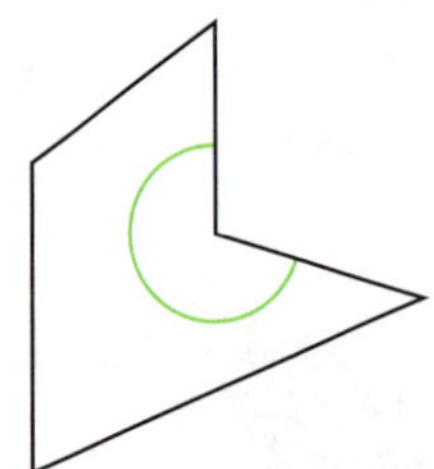

▲ *Concave shapes have at least one interior angle that is reflex (more than 180°).*

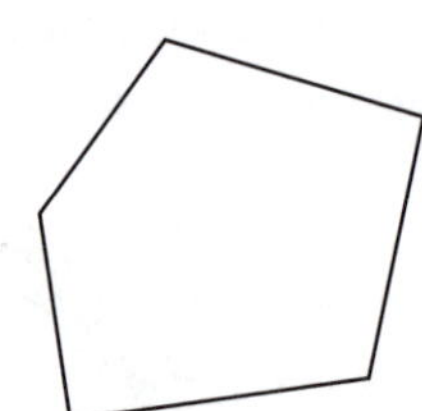

▲ *Convex shapes have no interior angles that are reflex.*

calculating angles

A distinction can be made between measuring angles and calculating angles. Measuring angles involves a tool of some sort, like a protractor. Calculating, however, relies more on reason and logic, skills that can take many years to develop. Although these activities are written for older students, some younger students may appreciate the challenge they provide.

1. Interior Angles in a Triangle

Preparation

If necessary, copy Blackline Master 15 onto overhead transparency sheets so that every student has a protractor.

Activity

1. Direct the students to cut a sheet of paper in half so that they have two smaller sheets and draw a large triangle on one sheet. Have them cut out the triangle and mark the three internal angles with arcs and the letters a, b, and c.

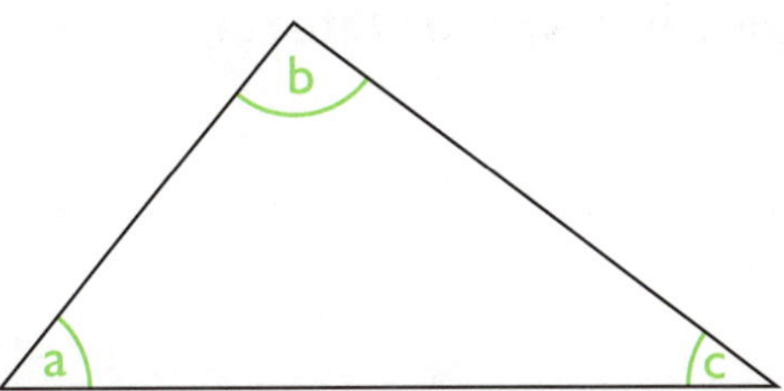

▲ *The students mark and label each interior angle in their triangles.*

2. On the second sheet of paper, ask the students to rule a straight line. Review what the students know about half angles. Discuss other names for them, such as "straight" angles and point out that a picture of the two angle arms making a half angle looks just like a straight line. Each student can draw a point on their line to represent the vertex.

▲ *Have the students mark a point on a line to indicate the vertex.*

Materials

- Protractor — 1 for each student (or use Blackline Master 15 on page 73)
- Scissors — 1 pair for each student

Did You Know?

The "vertex" is the point where two or more line segments join or intersect on a 2D shape, where three or more edges meet on a 3D shape, or the point where two or more angle arms intersect.

Angle arms are two line segments, real or imaginary, that share a common vertex and create an angle.

A half angle is one-half of a complete turn around a point. It is equal to 180°.

Did You Know?

Materials

- Protractor — 1 for each student (or use Blackline Master 15 on page 73)
- Scissors — 1 pair for each student

Did You Know?

3. Invite volunteers to identify the vertices of their triangles. Then, have all students tear off the corners of their triangles. The pieces should be quite large. Direct each student to arrange their triangle corners on one side of the straight line so that the vertex of each corner touches the point drawn on the straight line.

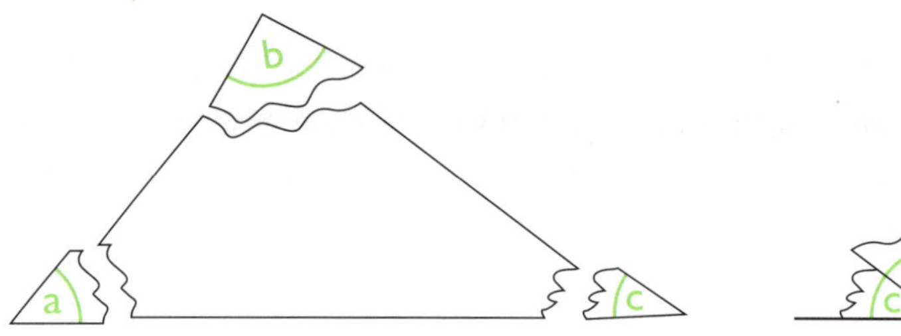

▲ *Invite the students to tear off and rearrange the corners of their triangles to form a half angle.*

4. Have each student compare his or her results with those of a partner. Prompt the students to talk about what they can conclude from the arrangement of angles. Ask, *When you combined the angles of your triangle, what type of new angle did you make?* (A half angle.) *How many degrees are in a half angle?* (180°.) *What does that tell you about the sum of the internal angles of a triangle?* (The sum of the internal angles of a triangle is equal to 180°.) Have the students measure each angle with a protractor to confirm their conclusions.

2. Interior Angles in a Quadrilateral

Preparation

If necessary, copy Blackline Master 15 onto overhead transparency sheets so that every student has a protractor.

Activity

1. As with the previous activity, the students can investigate the sum of the internal angles of a convex quadrilateral. This time, however, they will arrange the corners just around a point, not a line. The students will reach the conclusion that, because the four internal angles of a quadrilateral can be combined to form a full angle, the sum of the internal angles is equal to 360°.

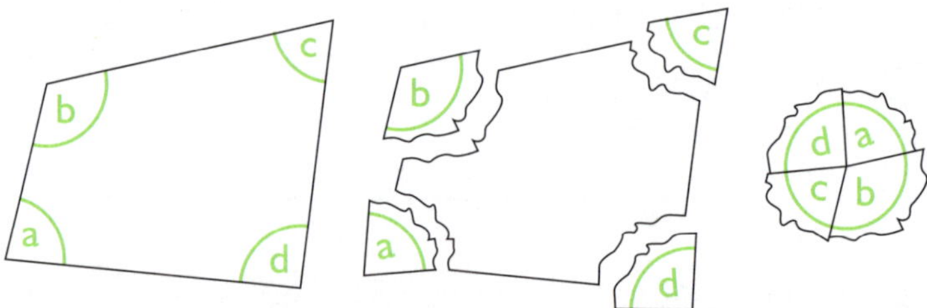

▲ *The corners of any quadrilateral can be torn off and rearranged to form a full angle.*

2. Challenge the students to use a similar method to investigate convex pentagons or hexagons. The students will find that the torn corners start to overlap. For instance, the torn corners of a convex hexagon will form 360° twice.

3. Pattern Block Angles

Preparation

No preparation is required.

Activity

1. The students can use a number of methods to calculate the value of the angles of pattern blocks. Challenge the students to begin with the angles of only one shape and use that shape to prove the value of all the angles of the other pattern blocks.

2. Say, *The sum of the interior angles of a triangle is equal to 180°.* If the students have trouble getting started, lead them through the following sequence:
   - *The triangle pattern block can be rotated, one corner at a time, over another triangle and each angle will align exactly. This means that the sum of the internal angles can be divided into three equal parts and that each angle must be 60° (180° divided by three angles).*
   - *The acute angles of two pale rhombuses can be aligned with the corner of a triangle pattern block. This means that each acute angle of a rhombus is 30° (60° angle of triangle divided by two acute angles of rhombuses).*

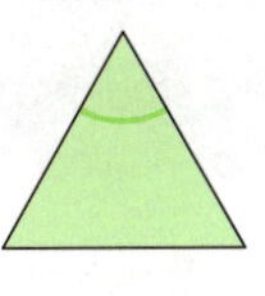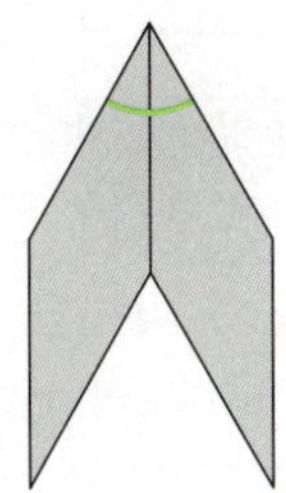

▲ *Guide the students to see that the acute angle of a pale rhombus pattern block is half the magnitude of an interior angle of the triangle.*

3. By placing "known" shapes on top of or beside "unknown" shapes, the students can continue to calculate all of the angles of the remaining shapes. Discuss how the students solved the problems and be sure to identify and challenge any assumptions that the students have not "proved". Encourage the students to develop their reasoning as much as possible, using words, pictures, or materials to explain their answers.

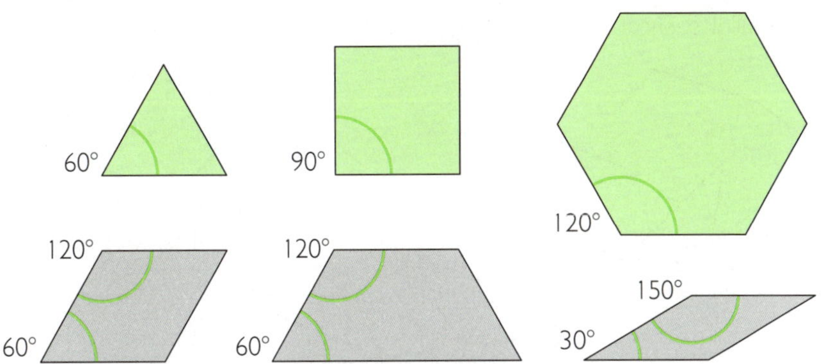

▲ *Using logic, the students will be able to calculate the angle of each pattern block corner.*

Materials

- Pattern blocks — 2 of each shape for each student or pair of students

Did You Know?

An acute angle is an angle that is less than 90°.

An "interior angle is an angle inside the perimeter of a polygon. Its angle arms are two adjacent sides of the polygon.

■ Materials

- Protractor — 1 for each student (or use Blackline Master 15 on page 73)

geo For more activities on calculating the magnitude of angles without a protractor, see *Paper Polygons*.

4. Triangulation

Preparation

If necessary, copy Blackline Master 15 onto overhead transparency sheets so that every student has a protractor.

Activity

1. Say, *One method used to calculate the sum of the interior angles of any polygon with more than three sides is "triangulation".* Direct the students to draw an irregular pentagon.

2. Have the students draw two non-intersecting diagonal lines inside their pentagons. Explain that the lines should not cross over and must start and finish at the vertices of the pentagon.

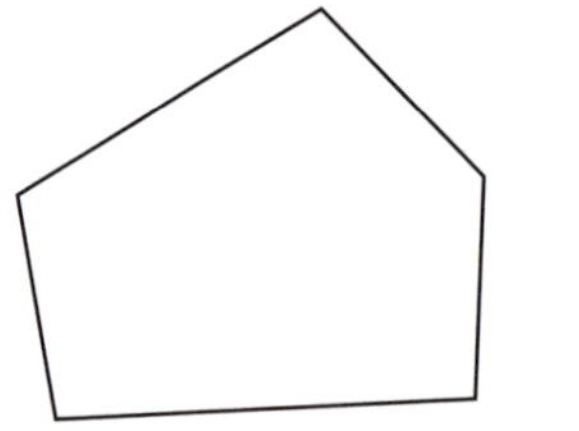 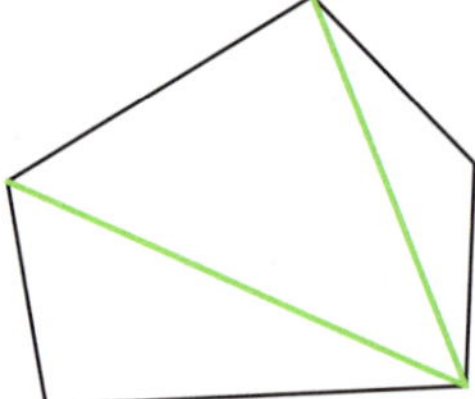

▲ *By dividing a shape into triangles, students can calculate the sum of the interior angles.*

3. Ask, *What do you notice about the shapes inside the pentagon?* (There are three triangles.) *Mark the internal angles of each of the triangles. Now use a different color of pencil and mark the internal angles of the pentagon. What do you notice?* (A combination of triangle angles combine to form each pentagon angle.)

4. Ask, *What do you know about the sum of the internal angles of a triangle?* (They are equal to 180°.) *If all the pentagon angles are made up of the internal angles of three triangles, what can you say about the sum of the internal angles of a pentagon?* (It must be equal to three lots of 180°, which is 540°.) Have each student use a protractor to check the accuracy of this statement.

5. The students might like to try the procedure with other shapes. A challenge for the students may be to consider concave shapes such as the one shown below. To discover the triangles within this shape, the shape needs first to be broken into smaller sections.

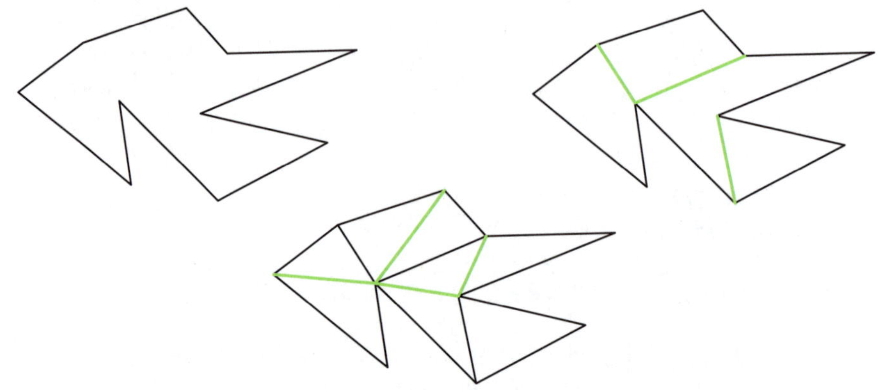

▲ *To use triangulation with unusual shapes, the students have to divide the shape into simpler sections.*

5. Triangulation Patterns

Preparation

No preparation is required.

Activity

1. As a follow-on from the previous activity, say, *Triangulation of pentagons, hexagons, and other shapes that don't have many sides is easy. How can we quickly calculate the sum of the internal angles of a polygon with a hundred sides?*

2. To help the students investigate this question, ask them to triangulate a number of different polygons and construct a table like the one below.

Name of polygon	Number of sides	Number of triangles formed	Sum of internal angles
Triangle	3	1	$1 \times 180° = 180°$
Quadrilateral	4	2	$2 \times 180° = 360°$
Pentagon	5	3	$3 \times 180° = 540°$
Hexagon	6	4	$4 \times 180° = 720°$
Heptagon	7	5	$5 \times 180° = 900°$
…	…	…	…
Hectagon	100	?	?

▲ *The students can construct a table to explore the patterns involved in triangulating polygons.*

3. As the pattern emerges, the students will see that the number of triangles formed is two less than the number of sides. If the pattern continues, the number of triangles formed in a hectagon must be 98. The sum of the internal angles is found by multiplying 180° by 98, which equals 17, 640°.

4. Guide the students in creating a general rule for finding the number of triangles found in a polygon of any number of sides (an "*n*-gon").

Name of polygon	Number of sides	Number of triangles formed	Sum of internal angles
Hectagon	100	$100 - 2 = 98$	$98 \times 180° = 17, 640°$
n-gon	n	$n - 2$	$(n - 2) \times 180°$

Materials

- Protractors from the previous activity

Did You Know?

A polygon is any simple two-dimensional closed shape formed by three or more straight line segments.

■ Materials

- Blackline Master 14 (page 72)
- Protractor — 1 for each student (or use Blackline Master 15 on page 73)

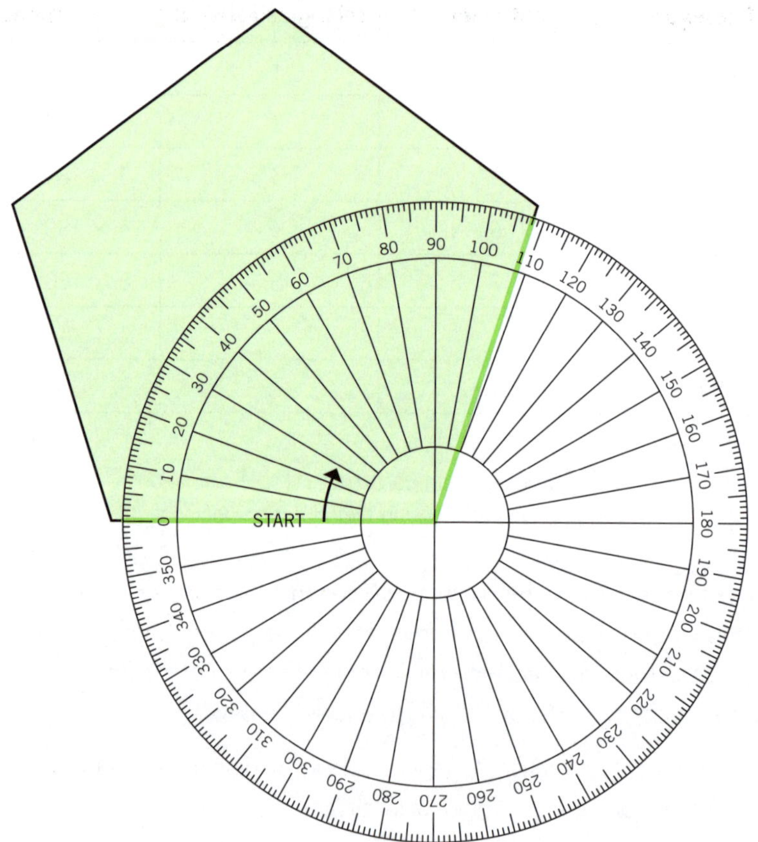

geo

See *Paper Polygons* for more activities about calculating angles without protractors.

6. Regular Polygons

Preparation

Make a copy of Blackline Master 14 for each student. If necessary, copy Blackline Master 15 onto overhead transparency sheets so that every student has a protractor.

Activity

1. Following on from the previous activities, explain to the students that if a polygon is regular, the size of each angle can also be calculated. Challenge the students to calculate a solution. For example, *A regular pentagon has five equal angles. The sum of the angles is 540°. 540° divided between five equal angles is 108°.*

2. Have the students check their calculations by measuring the interior angles of the regular polygons on Blackline Master 14 using a protractor.

▲ *Students use protractors to measure interior angles of regular polygons.*

A	B	C	D	E
F	G	H	I	J
K	L	M	N	O
P	Q	R	S	T
U	V	W		
X	Y	Z		

geo diagonal lines

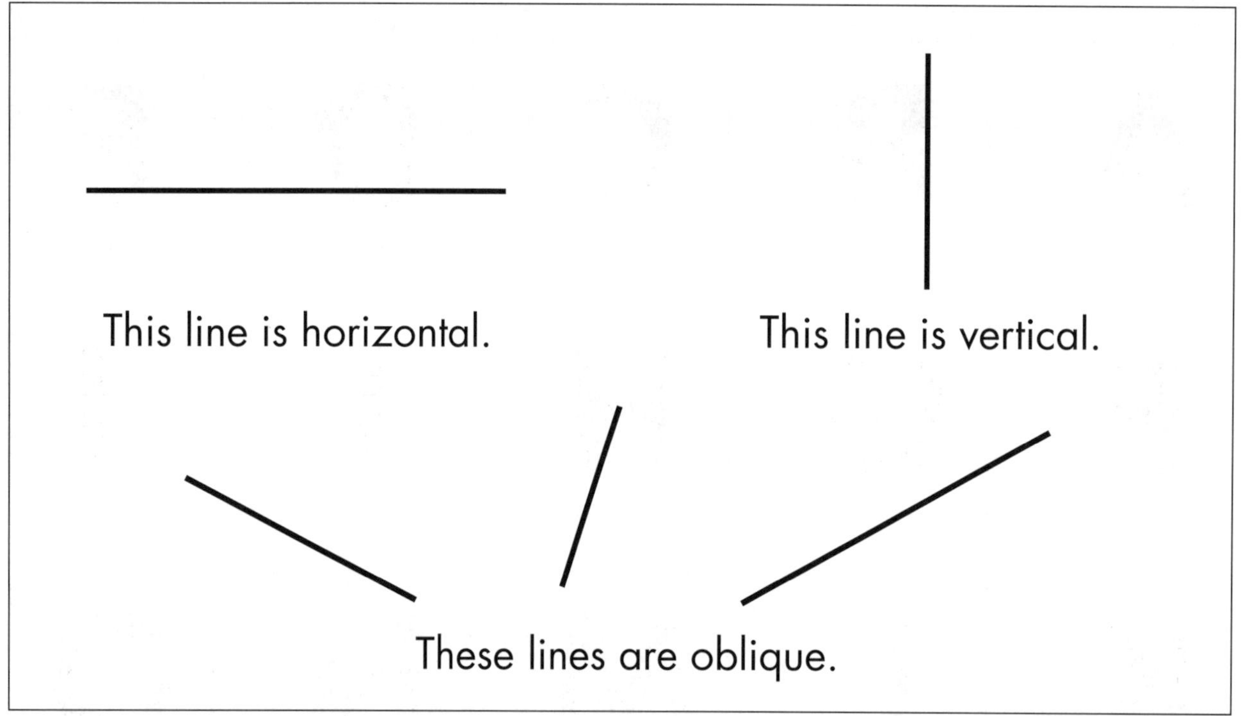

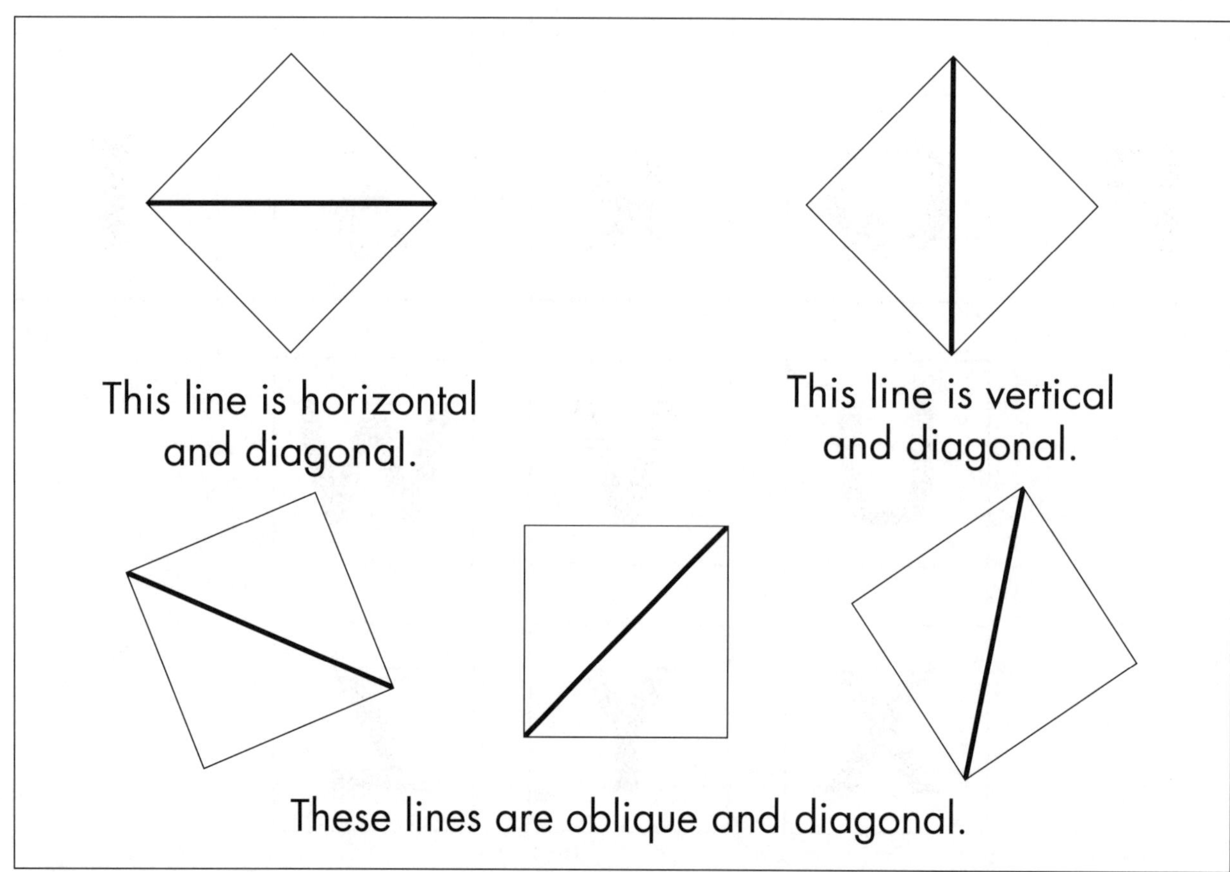

What is a diagonal line?

diagonals in quadrilaterals

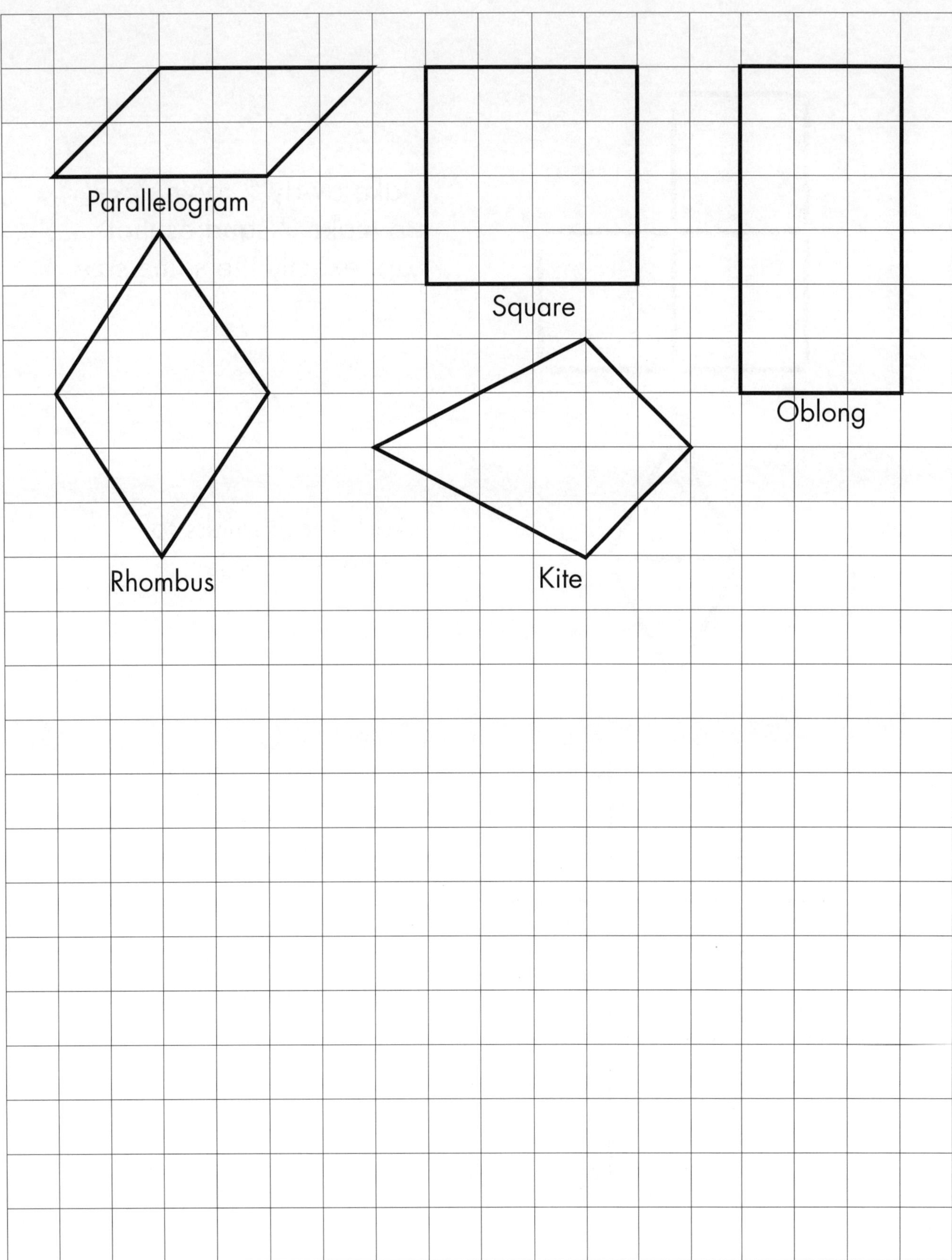

geo toothpick puzzles

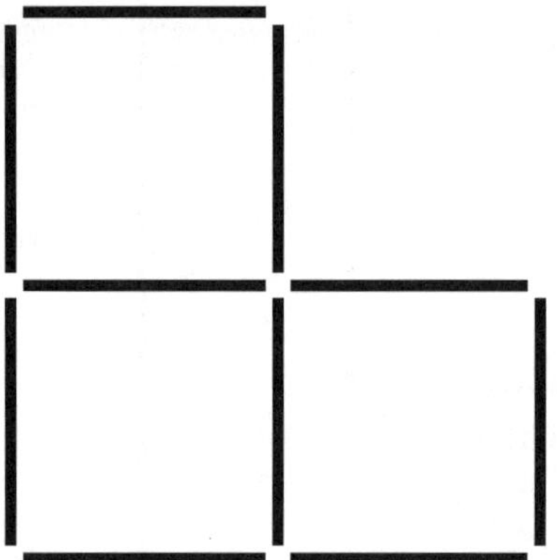

Take away 2 toothpicks to make 2 squares that are exactly the same size.

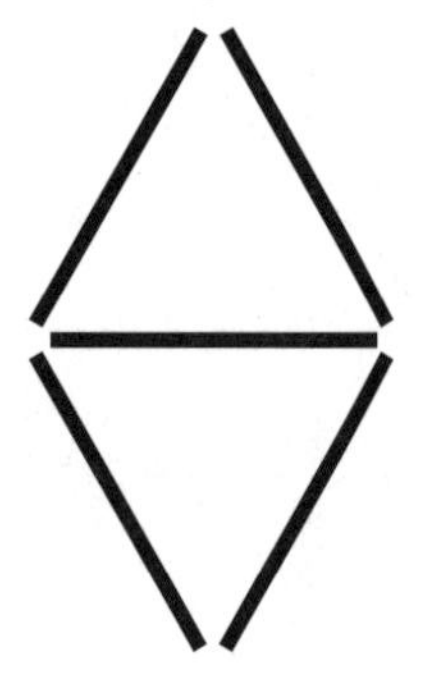

Add 4 toothpicks to make 5 triangles.

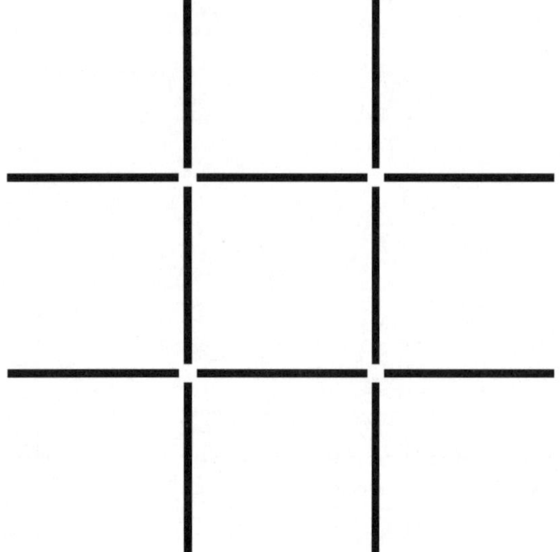

Move 3 toothpicks to make exactly 3 squares.

Blackline Master 4

geo
block writing

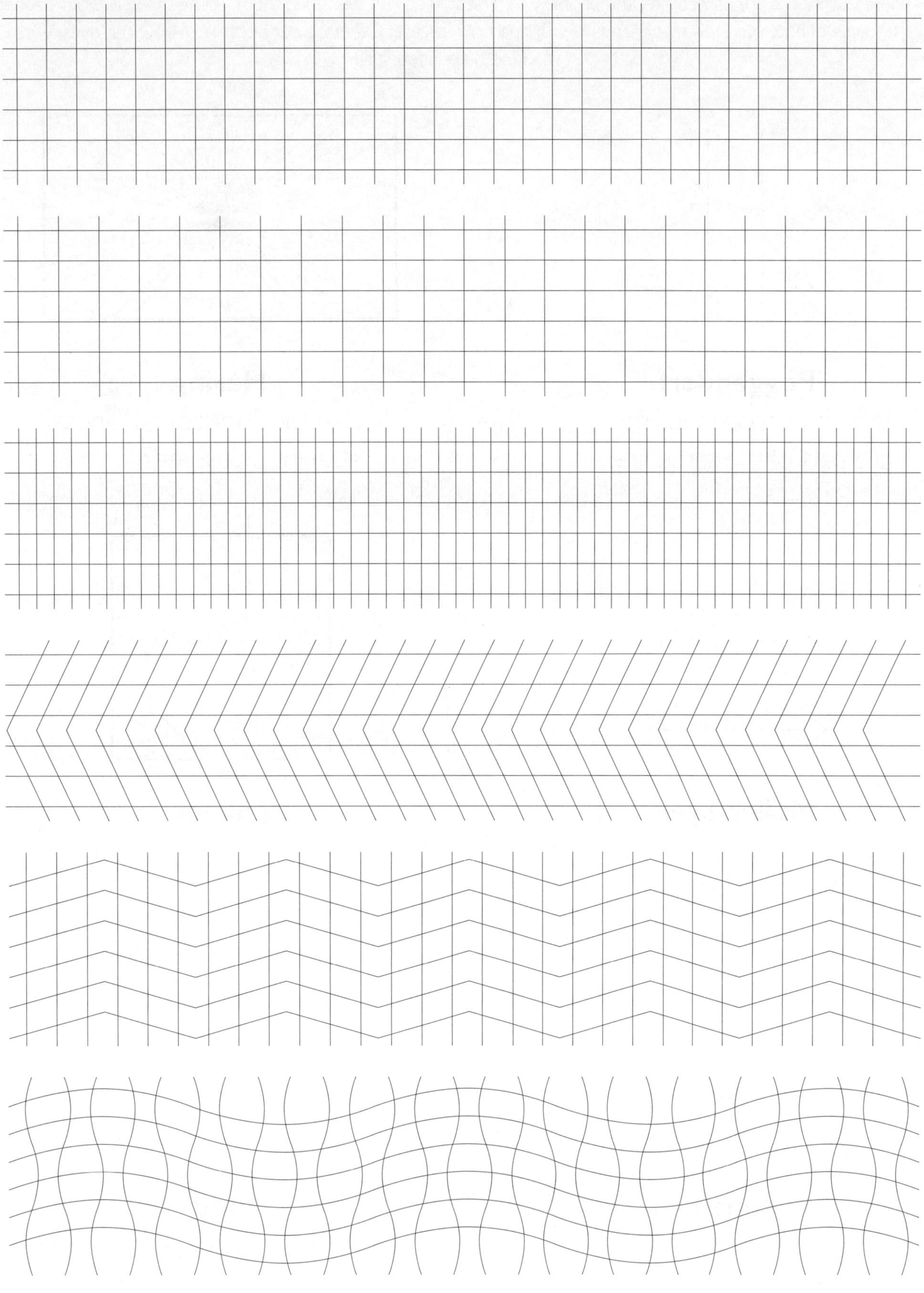

geo illusions

These illustrations are named after the people who first studied and described them. Some of the illusions happen because of how our brains work and how far apart objects are. Others occur because of how our eyes work. All of them are interesting!

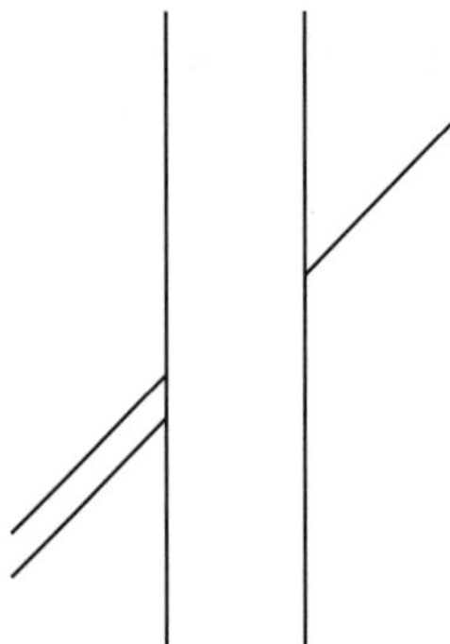

Poggendorf

Which two oblique lines are part of the same line?

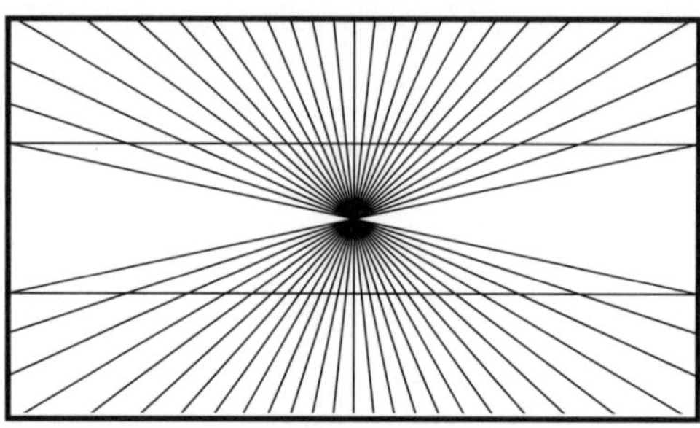

Hering

Are the "horizontal" lines straight or curved?

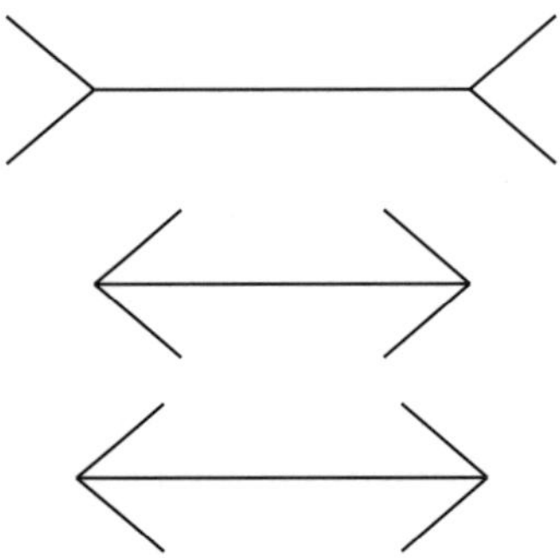

Müller-Lyer

Which two horizontal lines are the same length?

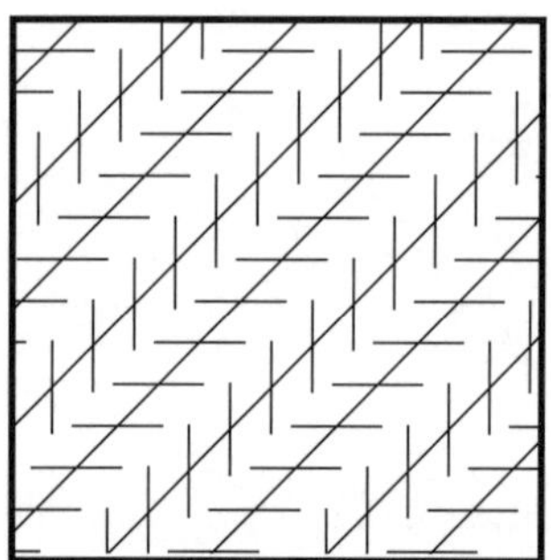

Zöllner

Are the oblique lines parallel?

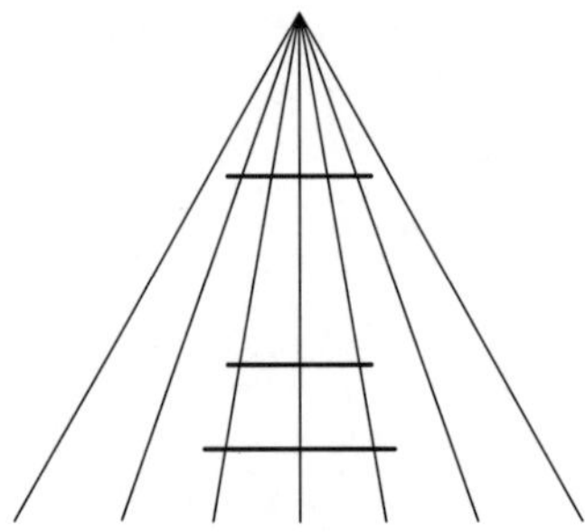

Ponzo

Which two horizontal lines are the same length?

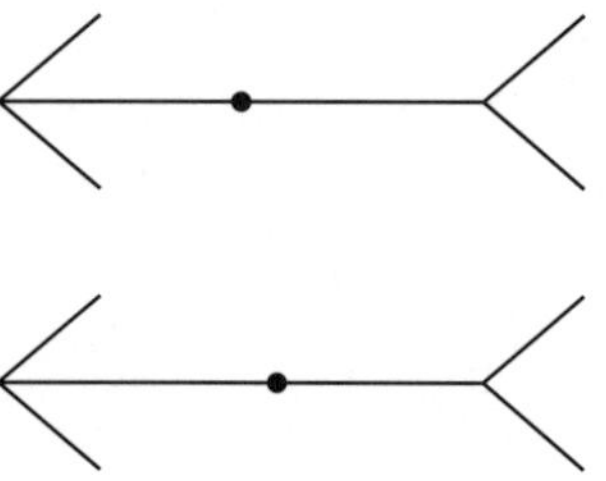

Judd

Which arrow has the dot halfway along the horizontal line?

Blackline Master 6

geo
circular dot paper

geo *triangular grid paper*

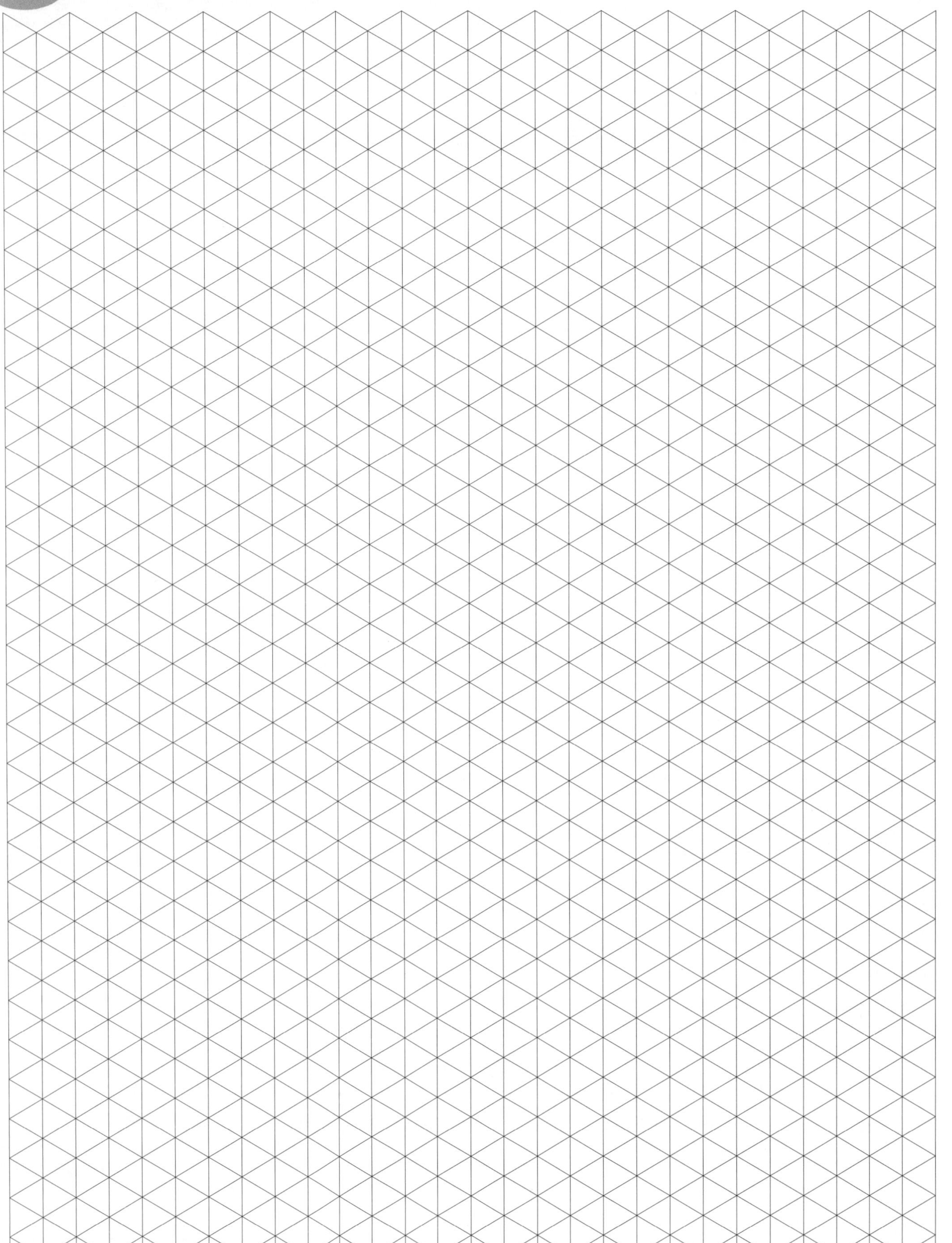

geo measuring interior angles

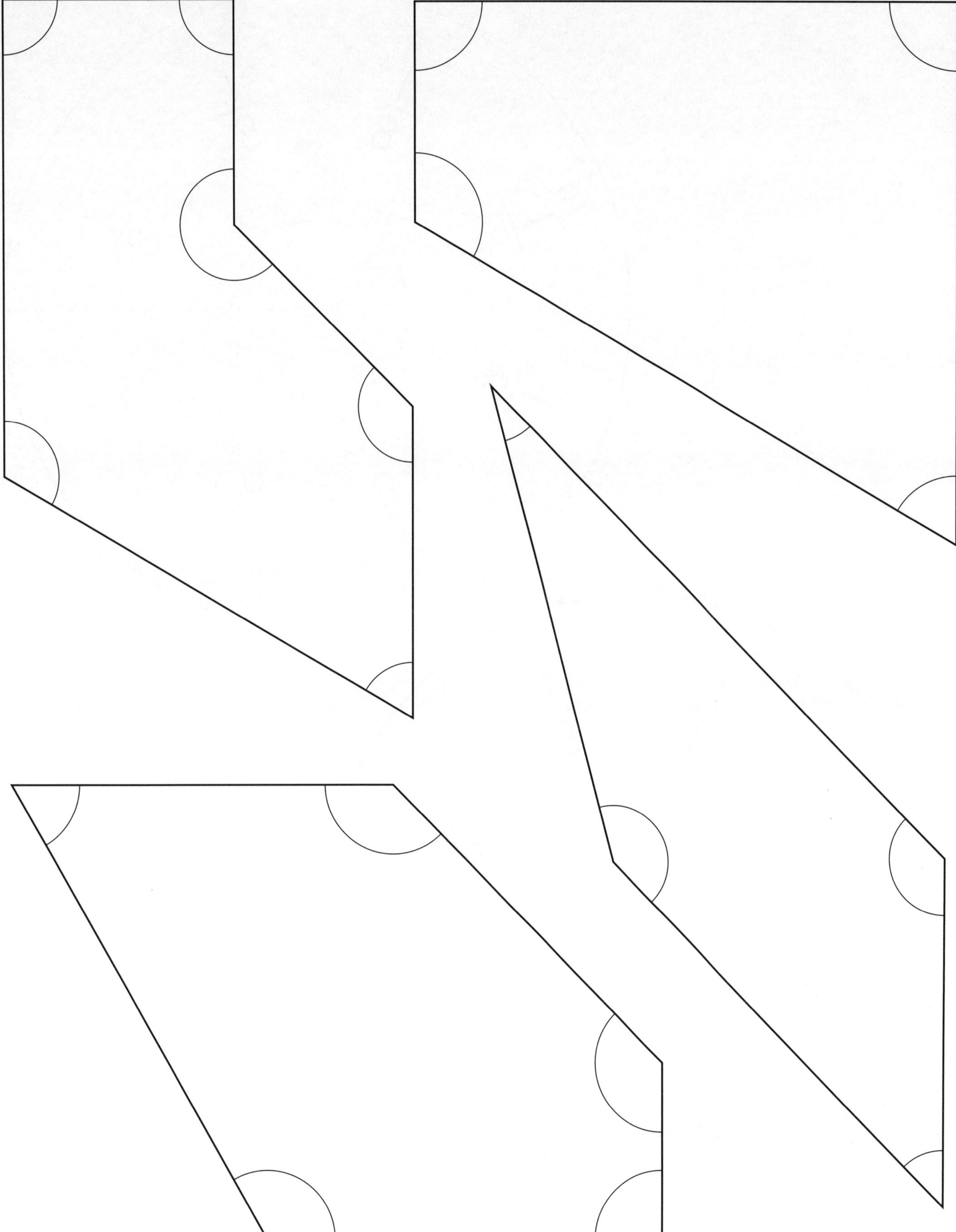

GEO All About Angles   67

eighth-angle tester

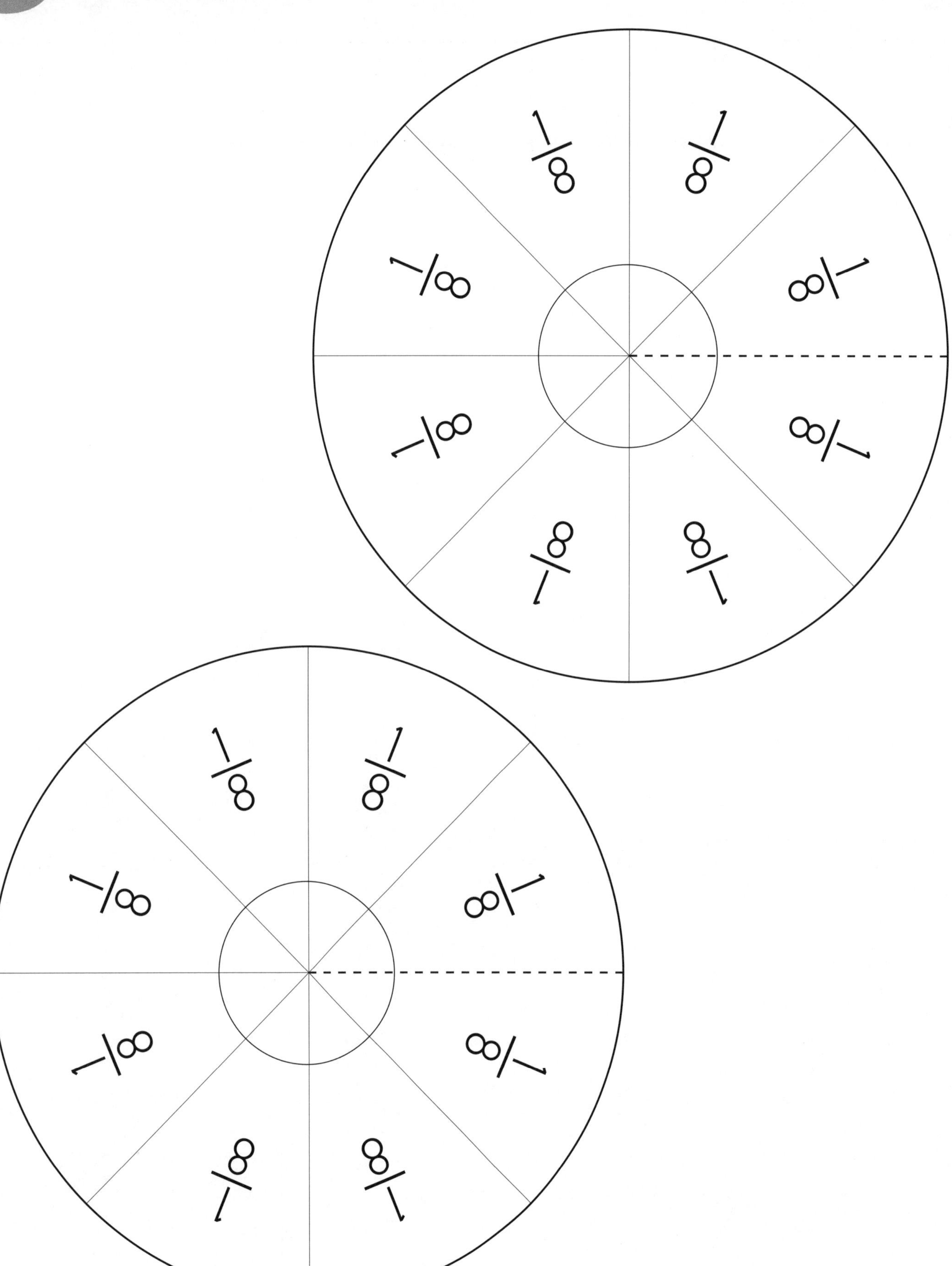

Blackline Master 10

geo
twelfth-angle tester

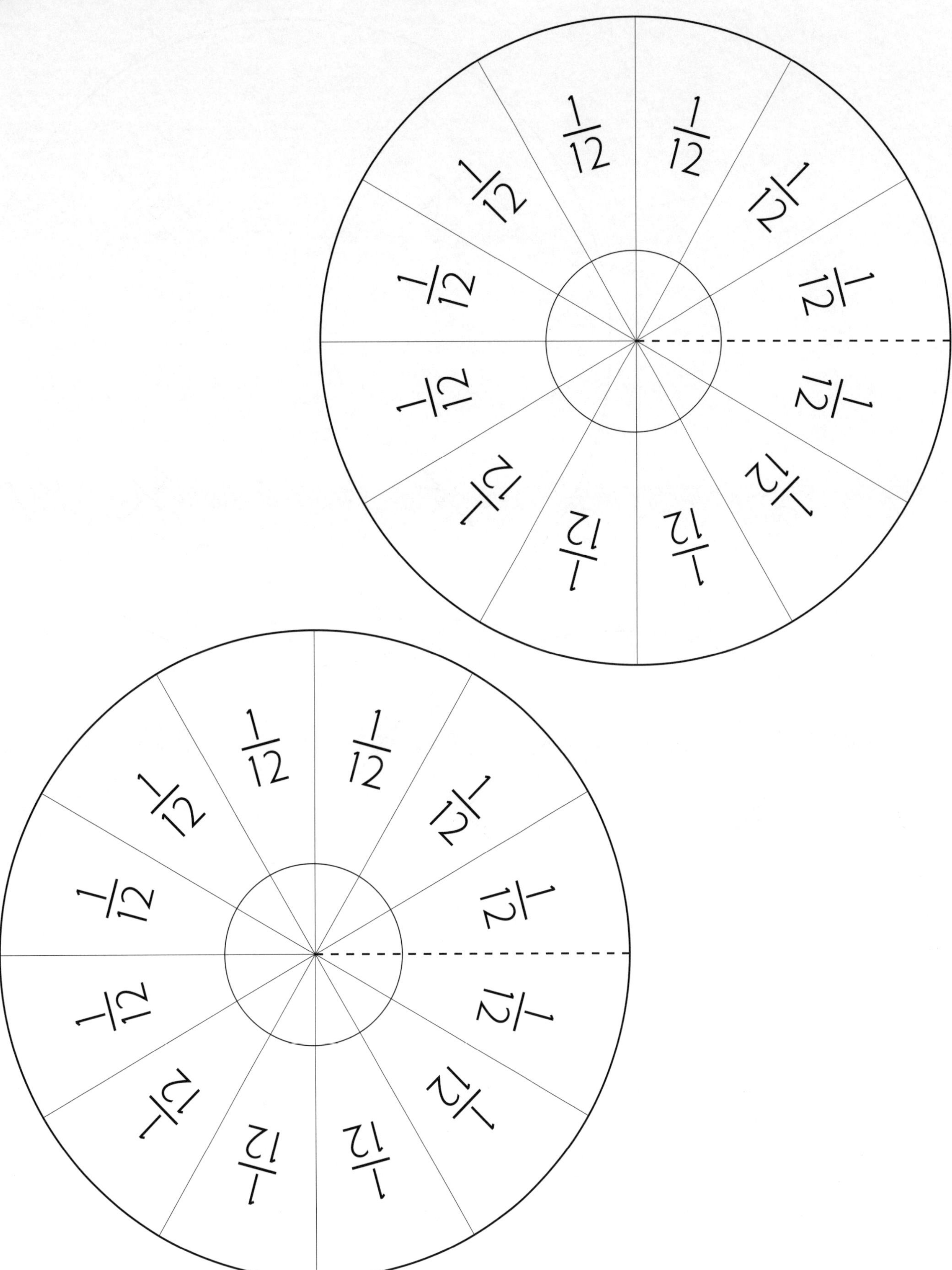

geo
quarter-angle tester

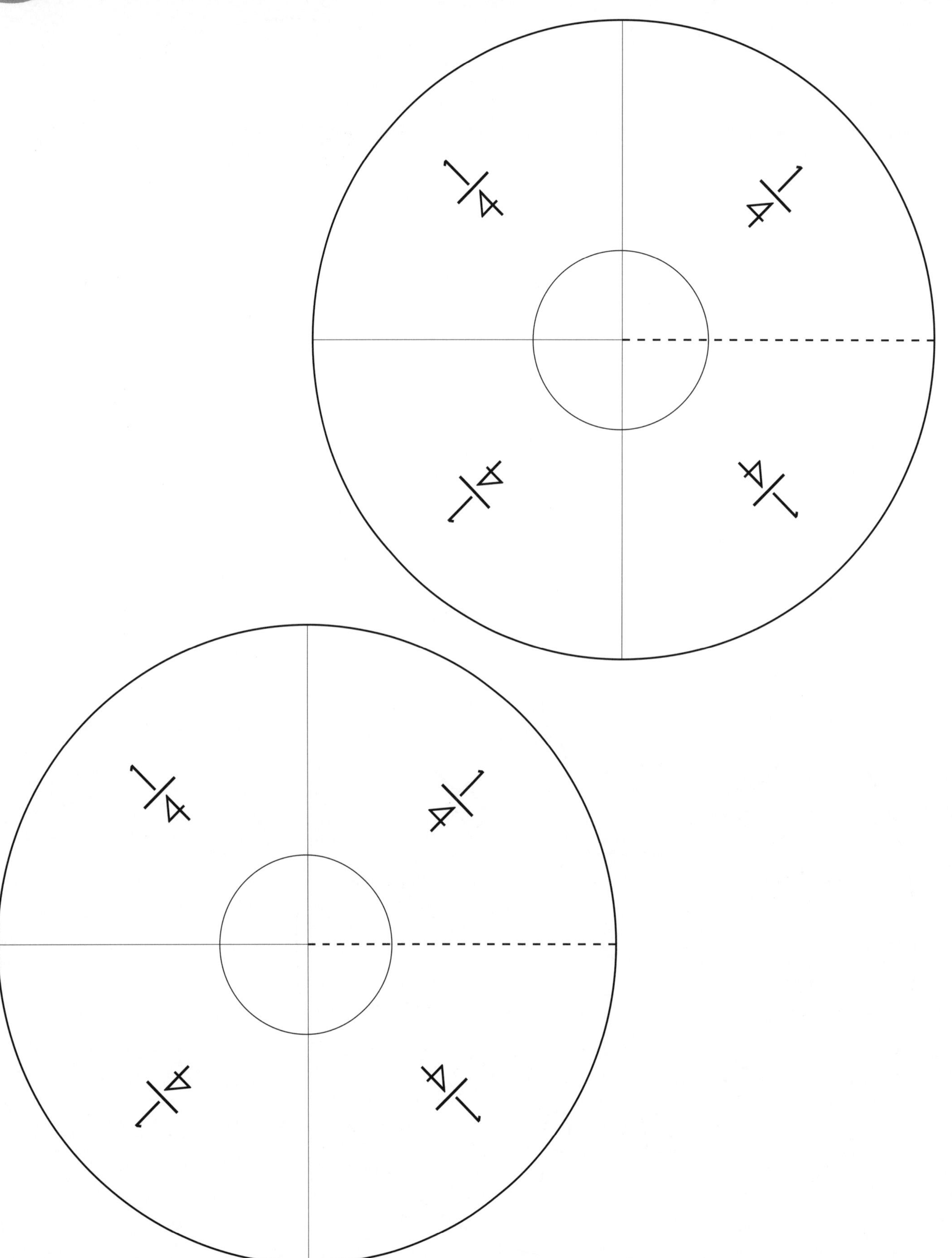

Blackline Master 12

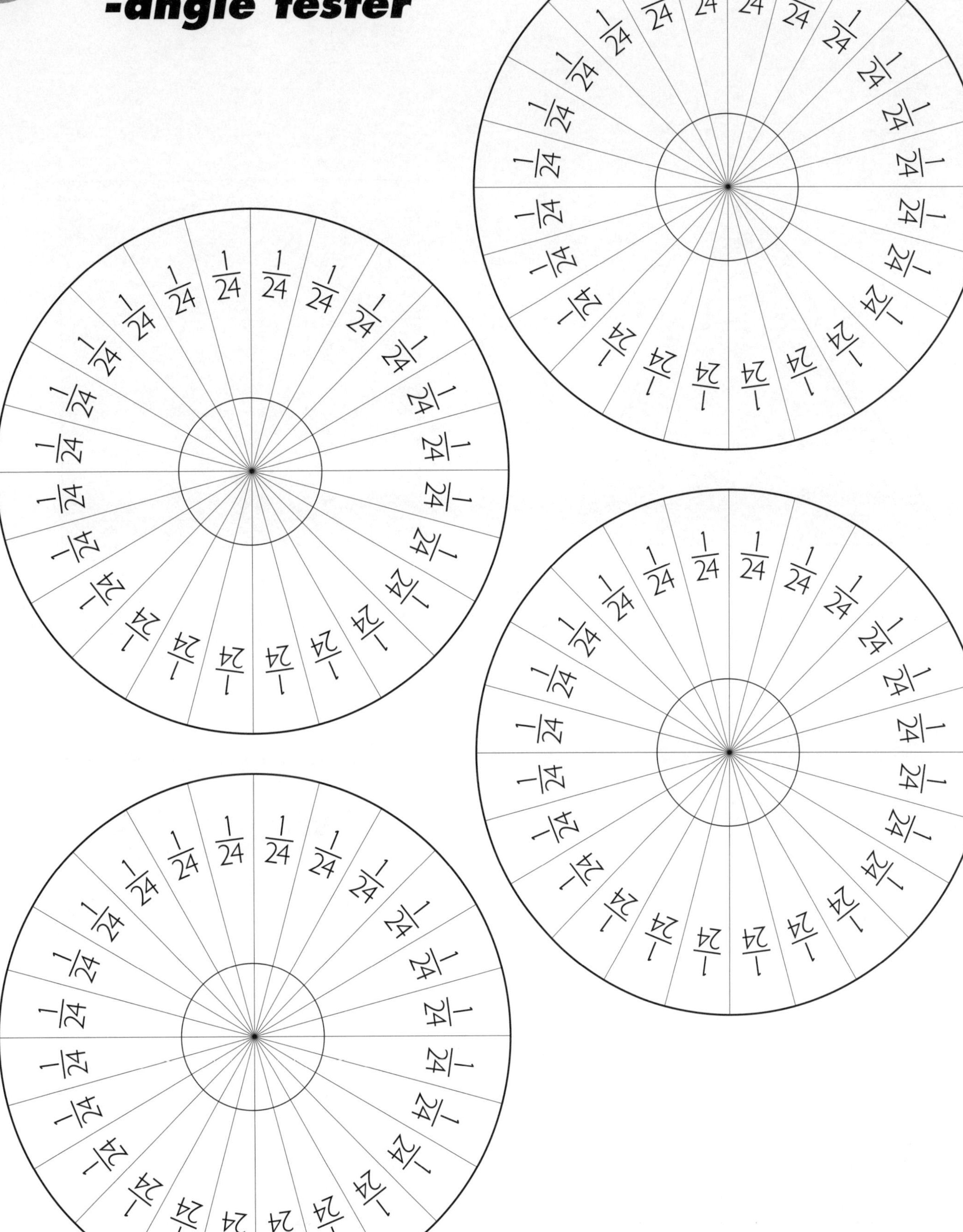

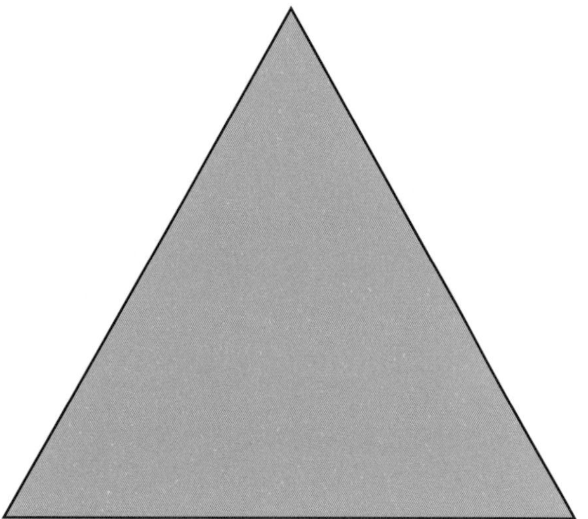

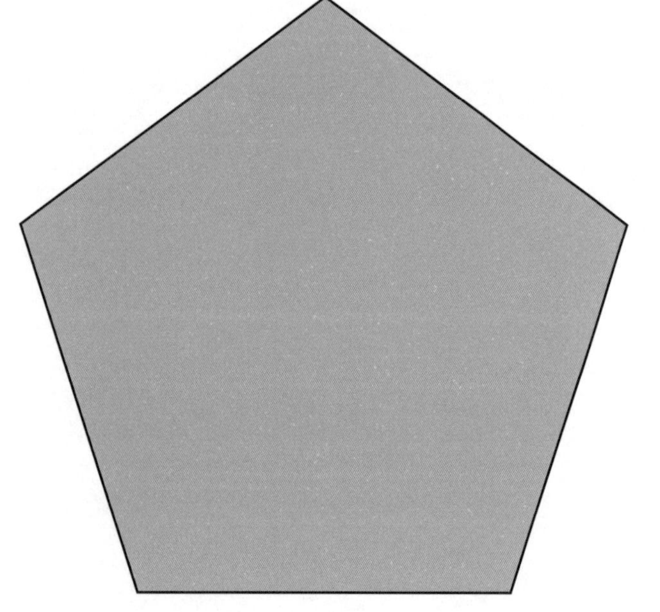

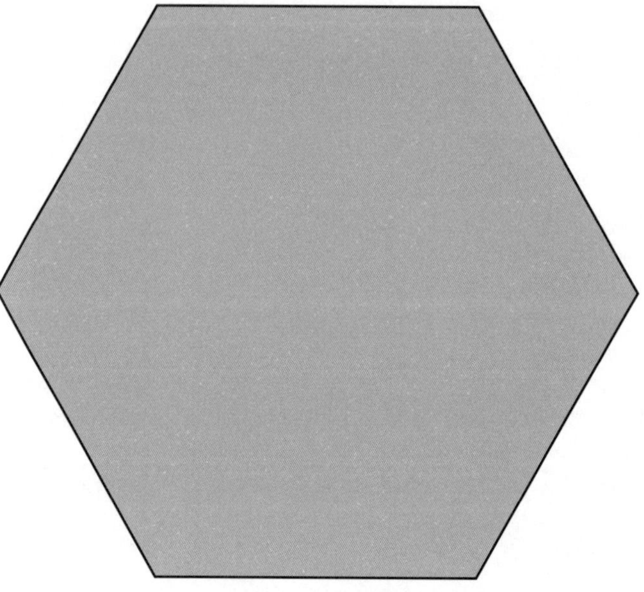

geo 360° protractors

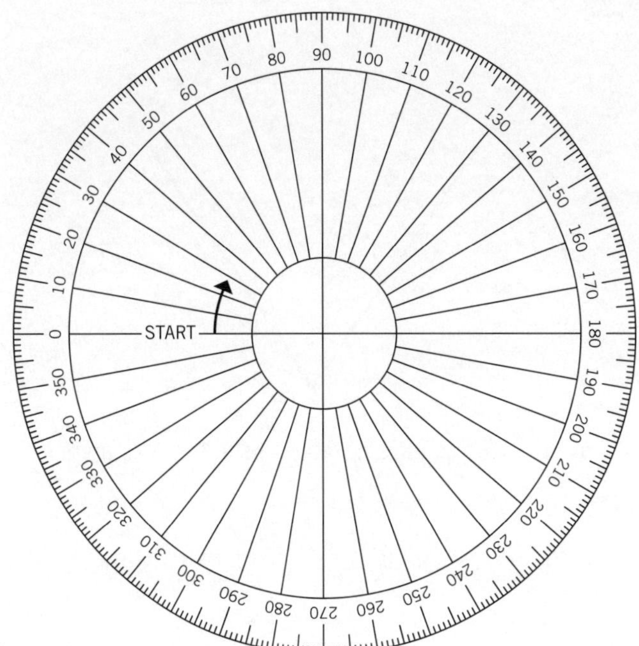

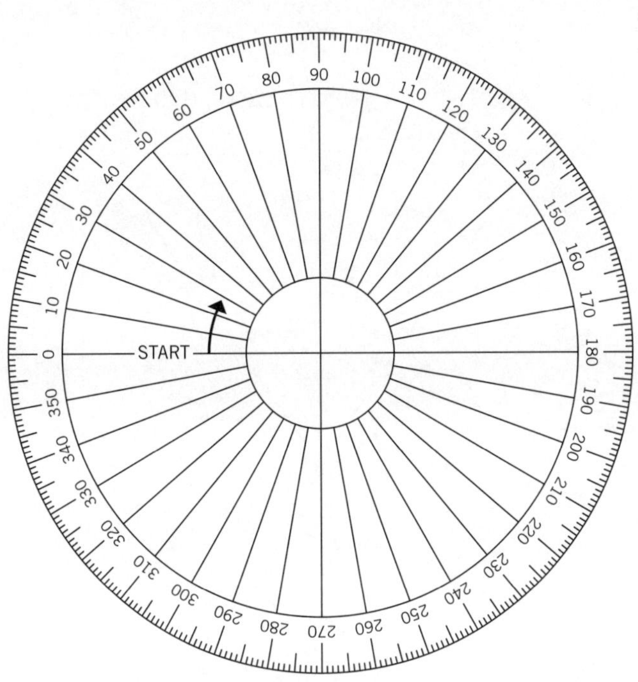

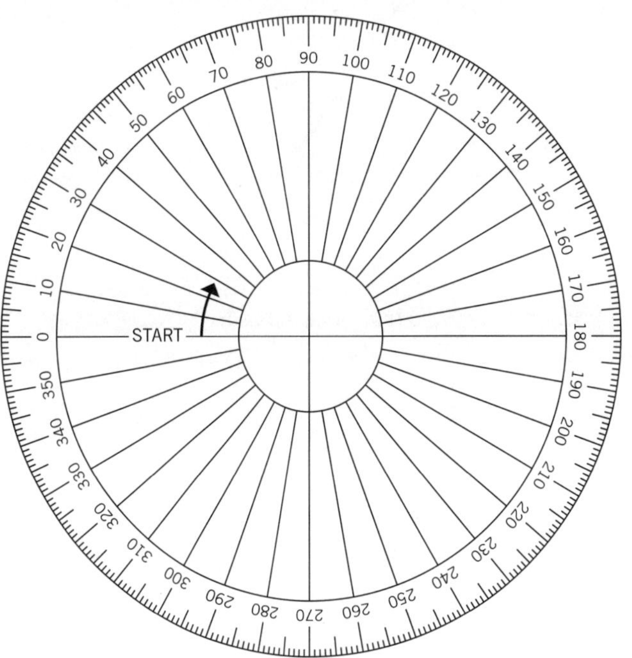

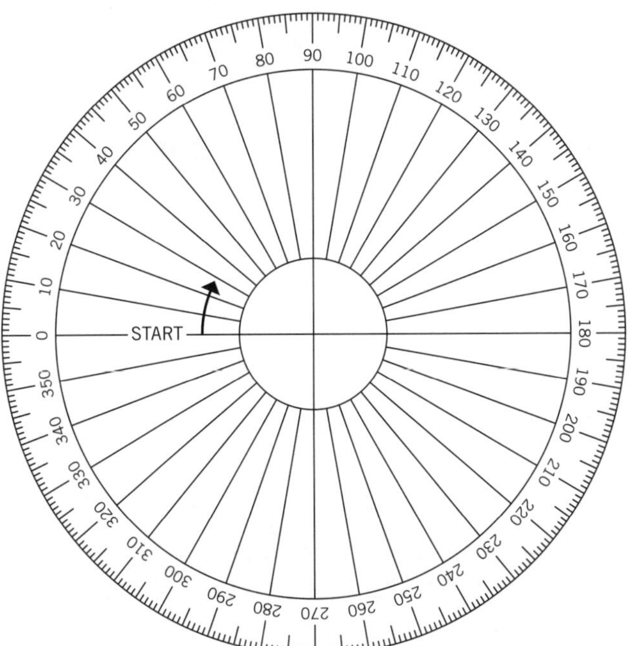

geo using a protractor

a. Identify the angle arms and vertex of your angle.

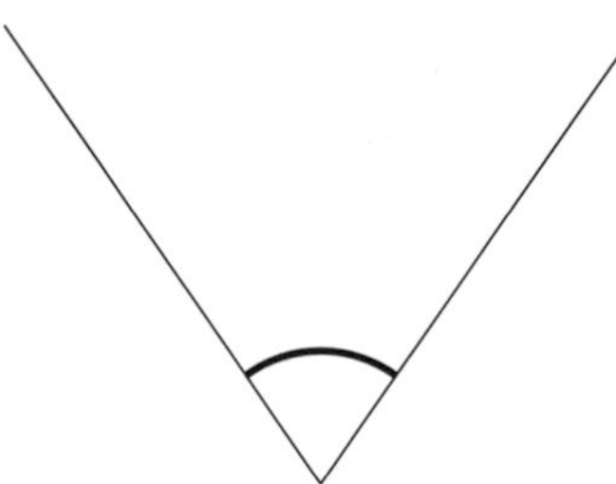

b. Choose which angle to measure. (Do the smaller one this first time.)

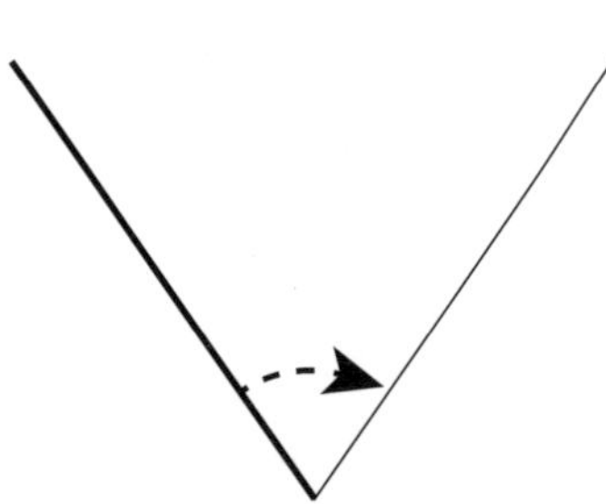

c. Imagine which angle arm has to move clockwise to the other to show the amount of turn.

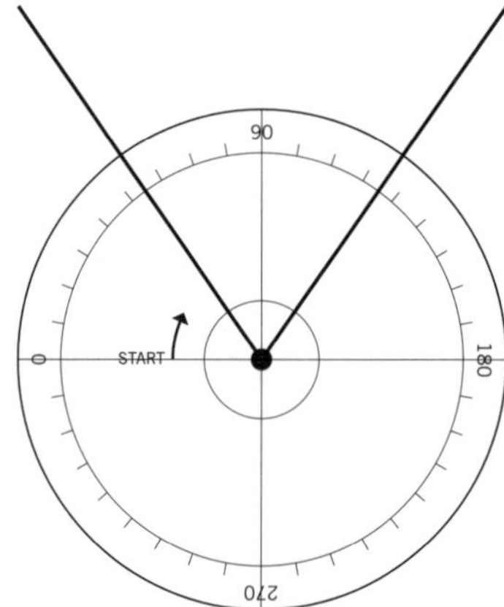

d. Place the centre of the protractor on the vertex of the angle.

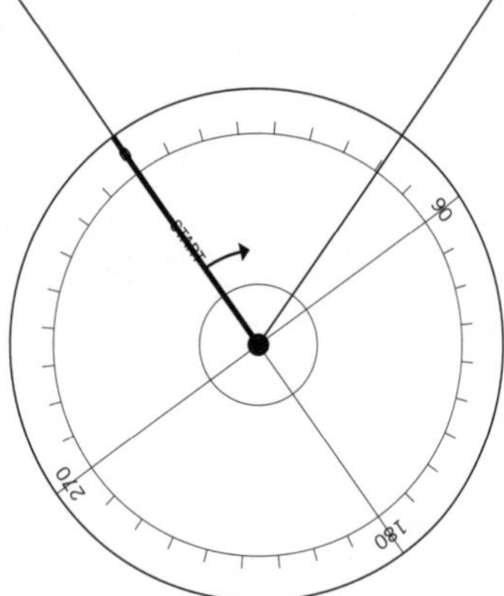

e. Place the protractor's "Start" line on the angle arm that you imagine moving to the position of other angle arm.

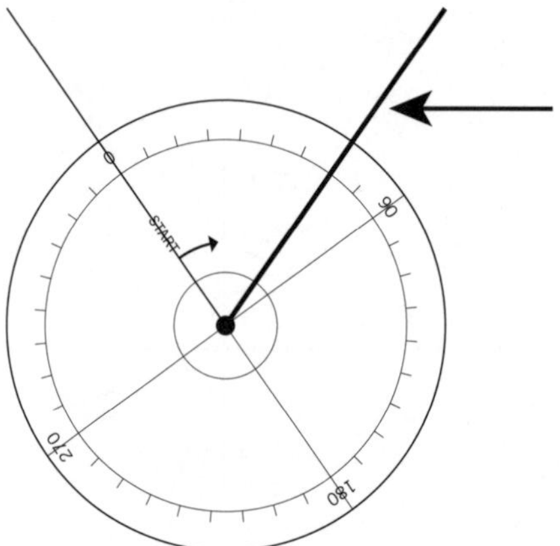

f. Find the protractor mark that lies on top of the second angle arm. This is the size of the angle in degrees.

geo angles in everyday life

Many words that we use to describe lines and angles have ancient Greek or Latin roots. Below are some of the more common words used to describe lines and angles with explanations of their meanings.

Angles

The word "angle" comes from a Latin word, *angulus*, which means "a little bending". Many of the other words used to describe angles also come from Latin.

- acute – Latin *acutus*, meaning "pointed" or "sharp" like a needle.

- obtuse – Latin *obtusus,* which means "dull" or "blunted".

- perigon – from two Greek words, *peri*, which means "around", and *gonus*, which means "angle" or "corner".

- reflex – Latin, *flectere*, which means "to bend".

- revolution – from the Latin word *volvere,* meaning "to roll" or "to turn".

- right – Old English *riht*, meaning "straight". A string with a weight on the end hangs straight down to form a right angle with level ground.

Points

The word "point" comes from the Latin word, *punctus*, "to puncture". A special type of point, a *vertex*, also has Latin origins. The word was used by the Romans and meant "whirl" or "eddy".

A word related to *vertex*, is *vertere*, meaning "to turn". It described what was thought to be the turning point of the heavens.

Types of Lines

The word "line" comes from the Latin word for flax, *linum*. The Romans used fibres from the flax plant to make linen thread.

- diagonal – from two Greek words, *dia*, meaning "across", and *gonus,* "angle" or "corner".

- parallel – from two Greek words, *para,* meaning "alongside", and *allenon,* which means "one another".

- perpendicular – Latin *pendere*, meaning "to hang".

Parts of a circle

Circles can be very interesting shapes to study. Below are just a few of the words that people use to describe the lines and regions found in them.

- arc – Latin *arcus*, referring to "bow", the weapon

- chord – Latin *chorda*, meaning "string"

- circle – Latin *circus*, "ring"

- circumference – Latin *circum*, "around", and *ferre*, "to carry"

- diameter – from two Greek words *dia*, "across", and *metron*, "a measure"

- radius – a Latin word that means "staff" or "rod"

triangles

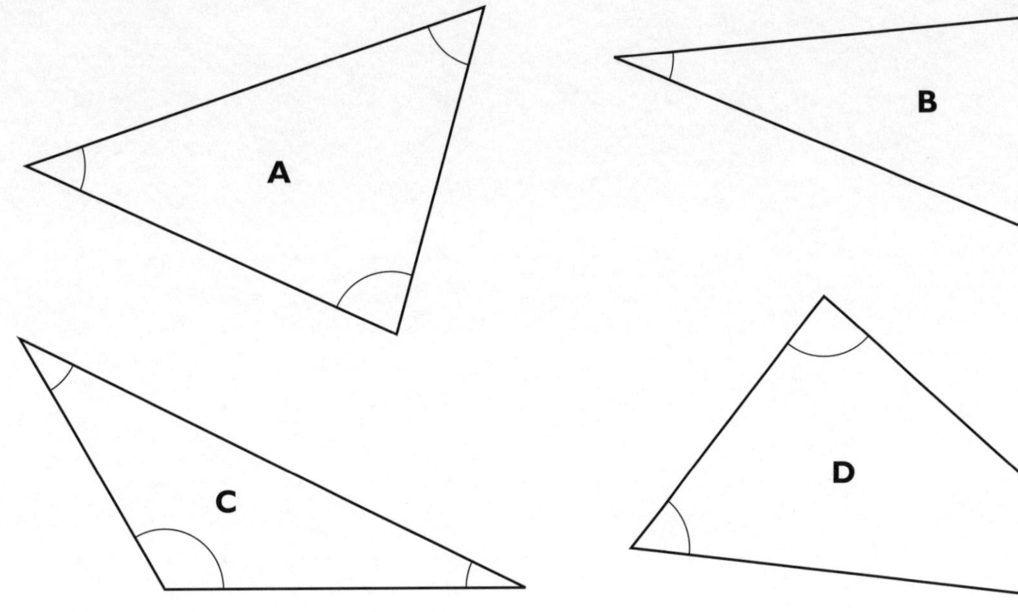

geo quadrilaterals

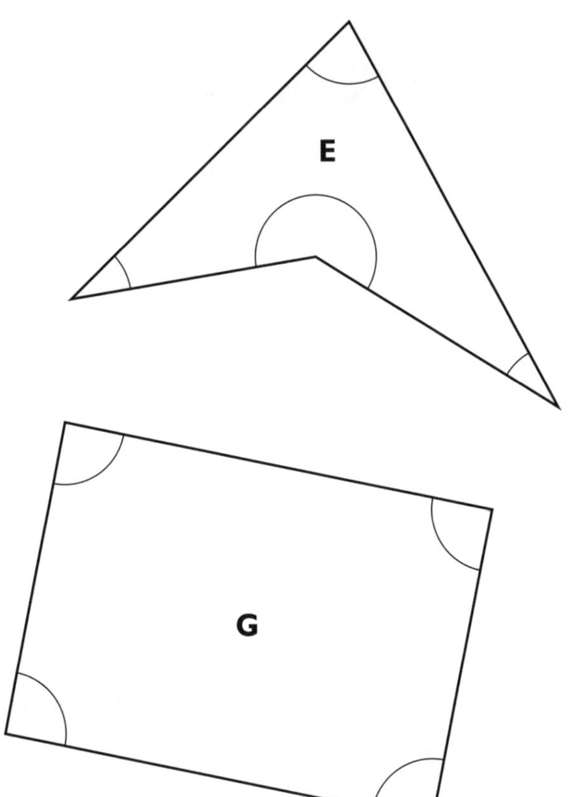

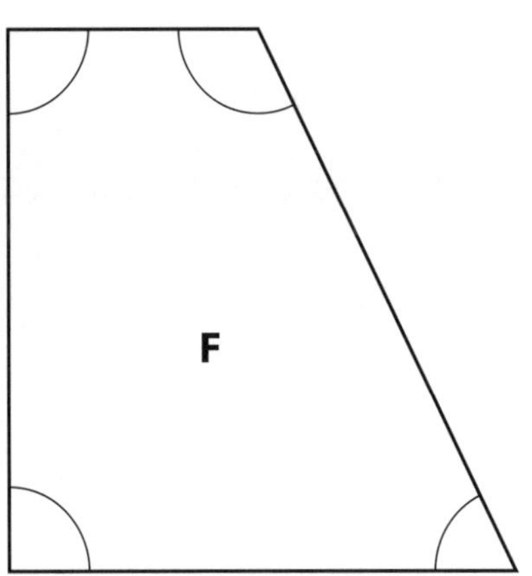

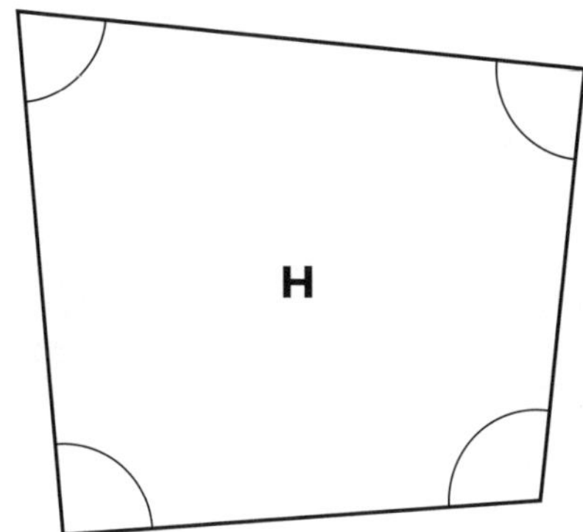

Roll 1 = acute angle

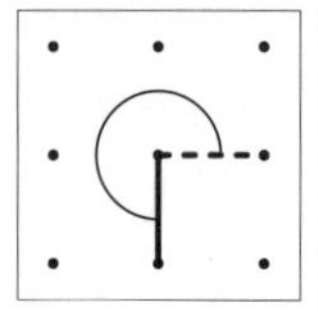

Roll 2 = right angle

Roll 3 = straight angle

Roll 4 = obtuse angle

Roll 5 = reflex angle

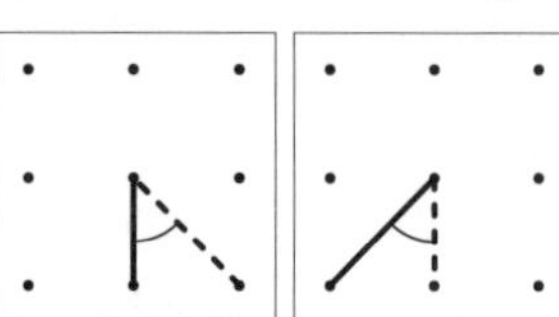
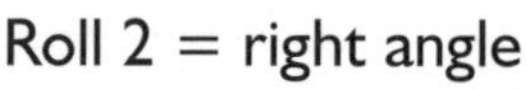
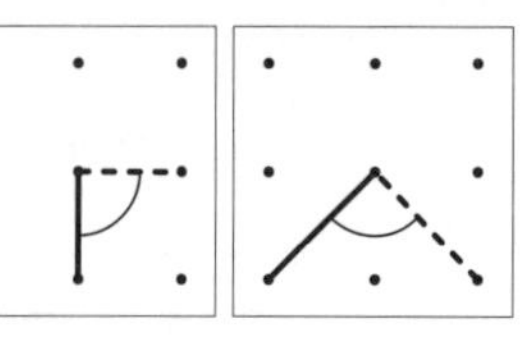
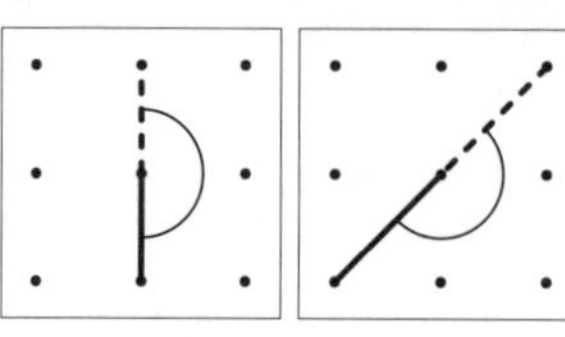
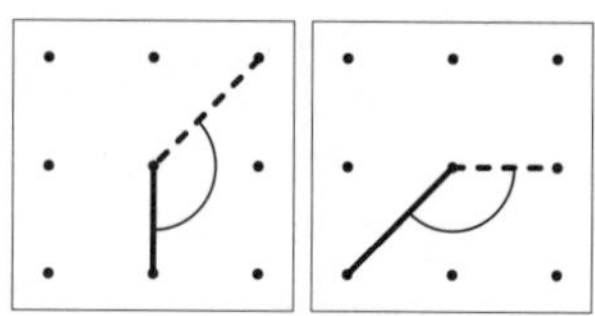

Roll 6 = free choice

~ anywhere on your trail ~

~ any type of angle ~

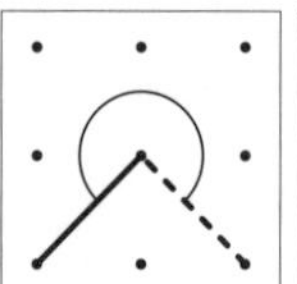
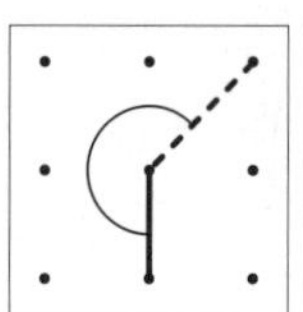
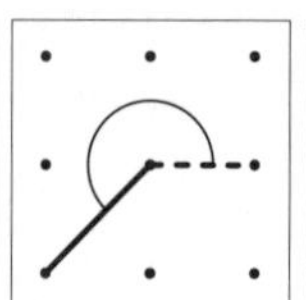
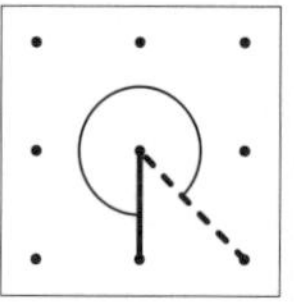
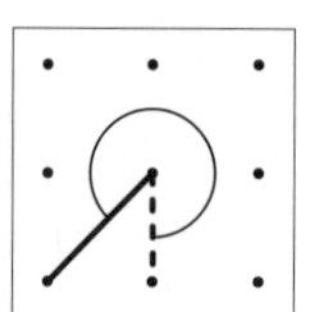

Blackline Master 20

glossary

Acute angle: An angle that is less than 90°.

Acute triangle: A triangle in which the largest angle is an acute angle.

Angle: When two line segments share the same vertex they create an angle. The size of the angle is determined by the amount of turn from one line segment (angle arm) to the other. See also dihedral angle.

Angle arms: Two line segments, real or imaginary, that share a common endpoint and create an angle.

Arc: A part of any curve. Usually used to describe part of the circumference of a circle.

Bisect: To bisect a line segment or angle is to divide in half.

Changeable angle: An angle in which the angle arms can be moved. Also known as dynamic.

Chord: A line segment that joins two points on the circumference of a circle.

Circumference: The boundary of a circle.

Concave: Concave 2D shapes have at least one interior angle that is larger than 180°. That is, they have a reflex angle.

Concentric: If two mathematically similar shapes have a common center they are said to be concentric.

Convex: The interior angles of a convex 2D shape are all less than 180°.

Degree: A unit of angle measurement. It is equal to 1/360 of one complete turn around a point. It is represented using the "°" symbol (e.g. "360° in a full turn").

Diagonal line: Any straight line segment joining two non-adjacent vertices in a polygon.

Diameter: Any line segment that passes through the center of a circle and joins two points on the circumference. A diameter is equal to two radii.

Dihedral angle: The angle between two adjacent faces of a three-dimensional shape.

Eighth angle: One-eighth of a complete turn around a point. It is equal to 45°.

Equilateral triangle: A triangle with three equal sides and angles. Each interior angle is 60°.

Exterior angle: An angle outside the perimeter of a polygon. Its angle arms are two adjacent sides of the polygon.

Fixed angle: An angle in which the angle arms do not move. Also known as static.

Full angle: One complete turn around a point. It is equal to 360°.

Full turn: See full angle.

Half angle: One-half of a complete turn around a point. It is equal to 180°.

Hexagon: Any polygon with six sides. A regular hexagon has six equal sides and each interior angle is 120°.

Horizontal: A line or surface that is completely level. A marble placed on a horizontal surface will not roll by itself.

Interior angle: An angle inside the perimeter of a polygon. Its angle arms are two adjacent sides of the polygon.

Irregular polygon: A polygon which does not have all sides equal or all angles equal.

Line: A straight line extends in either direction for an infinite distance into space. In everyday use, it usually refers to a line segment.

Line segment: A line segment is part of a straight line and has two end points. In everyday use, it is usually called a line.

Oblique: A line or surface which is neither vertical nor horizontal.

glossary

Oblong: A rectangle with adjacent sides of different lengths. Also known as a "non-square rectangle".

Obtuse angle: An angle greater than 90° but less than 180°.

Obtuse triangle: A triangle in which the largest angle is an obtuse angle.

Parallel: When two lines or planes are the same distance apart along their entire lengths, they are said to be parallel.

Parallelogram: A quadrilateral with opposite sides equal and parallel. Opposite angles are also equal.

Pentagon: Any polygon with five sides. A regular pentagon has five equal sides and each interior angle is 108°.

Perigon: See full angle.

Perpendicular: When two lines or surfaces intersect to form a right angle they are said to be perpendicular to each other.

Point: A position in space that is usually marked by a small dot.

Polygon: Any two-dimensional closed shape formed by three or more straight line segments.

Quadrant: A special sector of a circle. It is formed by an arc and two radii that form a right angle. It covers one quarter of the area bounded by a circle.

Quadrilateral: Any polygon with four sides.

Quarter angle: One-quarter of a complete turn around a point. It is equal to 90°.

Radius: A line segment that joins the center of a circle with any point on its circumference. The plural of radius is radii.

Ray: Similar to a line. It has one endpoint and extends infinitely in one direction. In everyday use, it is usually called a line.

Rectangle: A parallelogram with all angles equal to 90°.

Reflex angle: An angle that is more than 180° but less than 360°.

Regular polygon: A polygon with all angles and all sides equal.

Revolution: See full angle.

Rhombus: A parallelogram with four equal sides.

Right angle: An angle that is equal to 90°. See quarter turn.

Right triangle: A triangle in which the largest angle is a right angle.

Square: In geometry, a rectangle that has all sides equal in length.

Trapezoid/trapezium: A quadrilateral with exactly one pair of parallel sides that are of different lengths. In North America, it is known as a 'trapezoid'.

Triangle: Any polygon with three sides.

Twelfth angle: One-twelfth of a complete turn around a point. It is equal to 30°.

Twenty-fourth angle: One twenty-fourth of a complete turn around a point can be described as a twenty-fourth-angle. It is equal to 15°.

Vertex: The point where two or more line segments join or intersect on a 2D shape, or where three or more edges meet on a 3D shape. Also, the point where two or more angle arms intersect.

Vertical: A line or surface that forms a 90° angle with a horizontal line or surface.